TECHNISCHE UNIVERSITÄT BERGAKADEMIE FREIBERG

Die Ressourcenuniversität. Seit 1765.

CHAIR OF NUMERICAL
THERMO-FLUID DYNAMICS

Detailed Modeling of SI Engines in Fuel Consumption Simulations for Functional Analysis

Von der Fakultät für Maschinenbau, Verfahrens- und Energietechnik

der Technischen Universität Bergakademie Freiberg

genehmigte

DISSERTATION

zur Erlangung des akademischen Grades

Doktor-Ingenieur

(Dr.-Ing.)

vorgelegt von

Dipl.-Ing. Manuel Dorsch

geboren am 25.05.1986 in Esslingen am Neckar

1. Gutachter:	Prof. Dr.-Ing. Christian Hasse, Freiberg
2. Gutachter:	Prof. Dr.-Ing. Cornel Stan, Zwickau

Tag der Verteidigung: 22.04.2016

Bibliografische Information der Deutschen Nationalbibliothek

Die Deutsche Nationalbibliothek verzeichnet diese Publikation in der
Deutschen Nationalbibliografie; detaillierte bibliografische Daten sind
im Internet über http://dnb.d-nb.de abrufbar.

ISBN 978-3-8325-4270-2

Logos Verlag Berlin GmbH
Comeniushof, Gubener Str. 47,
10243 Berlin
Tel.: +49 (0)30 42 85 10 90
Fax: +49 (0)30 42 85 10 92
INTERNET: http://www.logos-verlag.de

ACKNOWLEDGMENT

I would like to acknowledge all who directly or indirectly supported me during my time at the powertrain simulation department of the BMW Group in Munich.

First and foremost, I would like to express my sincere gratitude to my thesis supervisor, Professor Christian Hasse, for all his helpful ideas and guidance along my studies.

My gratitude also extends Professor Cornel Stan for his time and supporting this thesis.

I would like to express my deepest appreciation towards my adviser at the BMW Group, Dr. Jens Neumann, for all his contributions of time and ideas that considerably improved the quality of this work. The joy and enthusiasm he has supporting his students was contagious and motivational for me.

I would like to acknowledge my mentor Dr. Florian Preuß for his assistance and providing a critical perspective in our constructive discussions.

My sincere thanks go to Luca Caberletti and Dominik Nebel from the development team of the high dynamic engine test bench at the BMW Group for helping me to obtain the experimental data. In particular, I would like to thank Tobias Billinger for his support and suggestions regarding emission calibration.

I would like to extend special thanks to the students Kevin Matros, Marco Römer and Stefan Frommater for their participation and contributions to my work. I also want to mention Ulrich Knoll and Dr. Sebastian Grasreiner for their valuable discussions and feedback.

Finally, and most importantly, I would like to thank my family who supported me in all my endeavors. I am especially grateful to one person above all others: Anni, I appreciate your years of patience and compromises, unconditional love, and catering service throughout this long journey.

"Vorne ist verdammt weit weg, wenn man ganz hinten steht."

(Frank-Markus Barwasser, German cabaret artist)

Dedicated to the memory of my mom
Birgit Dorsch

ABSTRACT

In order to manage increasingly complex combustion engines and to compensate the high costs of calibration work, a simulation environment for optimal parametrization of the engine management system is required. The aim of this thesis is to establish a coupled modeling approach combining the engine control and the vehicle powertrain with a detailed thermodynamic model of the combustion engine to simulate fuel consumption and in-cylinder gas emissions in various driving cycles. Therefore, a functional model of the electronic control unit is built with a modular architecture including all relevant modules. As fuel consumption specifications are often referred to a driving cycle, the dynamics of the powertrain and the vehicle are considered in a mechanical model. Linking these models with a crank-angle based combustion engine simulation opens up the possibility to support the development and calibration of future engines, demonstrated here for a turbo-charged spark ignited engine with direct injection and a fully-variable valvetrain. Thermodynamic processes are implemented within a 1D gas exchange model which allows to consider not only steady-state but also transient engine operation. The coupled system is extended by calculations of engine-out emissions considering the formation of nitrogen oxide, carbon monoxide, and hydrocarbons. Furthermore, tailpipe emissions are determined in an additional simulation model. The successful validation of this complex coupling technique is presented with exemplary results from all stages of the validation process. Finally, the advantage of this simulation methodology is shown by several application examples demonstrating the attained capabilities.

CONTENTS

LIST OF TABLES

NOMENCLATURE

$FMEP$ Friction Mean Effective Pressure

FTP Federal Test Procedure

FWD Front Wheel Drive

H_2O Water

HC Hydrocarbons

HIL Hardware-in-the-Loop

IVC Intake Valve Closing

IVO Intake Valve opening

MAF Mass Air Flow sensor

MT Manual Transmission

N_2 Nitrogen

$NDIR$ Nondispersive Infrared Detector

$NEDC$ New European Driving Cycle

NO Nitrogen Oxygen

NO_2 Nitrogen Dioxide

NO_x Nitrogen oxide

O_2 Oxygen

OBD On-Board Diagnostic

ODE Ordinary differential equation

Pd Palladium

$PEMS$ Portable Emission Measurement System

PI Proportional-Integral

PID Proportional-Integral-Derivative

PM Particulate Matter

PN Particle Number

Pt Platinum

PTA Pressure Trace Analysis

RCA Random Cycle Aggressive

RDE	Real Driving Emissions
Rh	Rhodium
RTW	Real-Time Workshop
SCR	Selective Catalytic Reduction
SI	Spark Ignition
SiL	Software in the Loop
SOC	Start Of Combustion
TCS	Traction Control System
TDC	Top Dead Center
TKE	Turbulent Kinetic Energy
TWC	Three-way Catalytic Converter
UDC	Urban Driving Cycle
USA	United States of America
VS	Vehicle Speed
VVT	Full Variable Valve Actuating

GREEK SYMBOLS

Symbol	Meaning	Unit
α	road slope	$^\circ$
α_{cyl}	in-cylinder heat transfer coefficient	$W/(m^2K)$
α_c	combustion conversion coefficient	$-$
α_{wall}	heat transfer coefficient for cylinder wall	$W/(m^2K)$
β_c	combustion form factor	$-$
δ_L	laminar flame thickness	m
δ_{oil}	oil film thickness	m
$\dot{\omega}_k$	gas phase reaction rate	$mole/(m^2s)$
η	efficiency factor	$-$
γ	geometric parameter	$-$
Γ_{Ce}	active site density of cerium	$kmol/m^2$

κ	isentropic exponent	—
λ	air-fuel equivalence ratio	—
λ_b	thermal conductivity	—
$\lambda_{friction}$	friction coefficient for fluids	—
λ_{tire}	tire-vehicle slip	—
μ	dynamic viscosity	kg/(ms)
μ_c	discharge coefficient	—
$\mu_{friction}$	friction coefficient for roads	—
μ_{oil}	dynamic viscosity of oil	kg/(ms)
μ_{TH}	discharge coefficient of the throttle	—
ν	kinematic viscosity	m^2s
ν_{mr}	stoichiometric coefficient of the reaction	—
ω	revolution speed	min^{-1}
ω_r	reaction rate	—
ϕ	crank angle	deg
ϕ_5	crank angle at 5% fuel conversion	deg
ϕ_{bd}	crank angle of period between 5-90% fuel conversion	deg
ψ	association factor for solvents	—
ρ	density	kg/m^3
τ	Taylor microscale	m
$\tau_{fuel,oil}$	penetration depth of fuel into the oil film	m
θ	surface coverage	—
φ	crank angle	deg
ς	Thiele modulus	—
ϑ	molar volume	—

LATIN SYMBOLS

Symbol	Meaning	Unit
$\dot{m}''$	convective mass flux	kg/h

$\dot{m}$	mass flow	kg/h
$\dot{Q}$	heat transfer	W/m^2
$\dot{s}_k$	surface reaction rate for species k	mole/(m^{2}s)
$\dot{v}$	acceleration	m/s^2
$[X]$	molar concentration	mole/m^3
A	Arrhenius parameter	cm^3/mole
a	parameter for precious metal loading	−
A, B, C	parameter to determine the dynamic oil viscosity	−
A_1	area behind the throttle	m^2
$A_{catalytic}$	catalytic active surface	m^2
A_{cs}	cross-sectional area	m^2
A_{cyl_head}	surface of cylinder head	m^2
A_{cyl_liner}	surface of cylinder liner	m^2
A_e	surface of entrained volume	m^2
A_{f_veh}	frontal area vehicle	m^2
A_{flame}	flame surface	m^2
A_{oil}	oil surface	m^2
A_{rep}	respective area	m^2
$A_{TH,max}$	maximum cross-sectional area of the throttle	m^2
A_{wall}	surface of cylinder walls	m^2
A_z	surface of the cylinder wall of zone z	m^2
AF	air flow	kg/h
B_clutch	bit clutch	−
$bmep$	break mean effective pressure	bar
c	coefficient	−
c_k	concentration profile of species k	−
$c_{p,b}$	specific heat capacity of the burned composition	J/(kgK)
c_p	specific heat capacity at constant pressure	J/(kgK)

$c_{quenching}$	factor for HC emissions due to quenching	−
c_v	specific heat capacity at constant volume	J/m^3K
CA	crank angle	deg
D	diffusion coefficient	m^2/s
d	diameter	m
$d_{quenching}$	quenching distance	m
E	energy	J
E_a	activation energy	cal/mole
EVT	valve timing exhaust valve	°CA
F	force	N
f_0, f_1, f_2	factors of the road load	$N, \frac{N}{km/h}, \frac{N}{(km/h)^2}$
F_{acc}	acceleration resistance	N
F_{aero}	aerodynamic force	N
$F_{cat,geo}$	catalytic surface factor	−
F_{load}	normal load	N
F_{roll}	rolling resistance	N
F_{slop}	slop resistance	N
$F_{x,gross}$	maximum longitudinal force	N
F_x	longitudinal force	N
g	gravity acceleration	m/s^2
g^*	mass transfer conductance	−
h	enthalpy	J/kg
IGN	ignition angle	°CA
IVT	valve timing intake valve	°CA
J	moment of inertia	kgm^2
$j_{k,a/r}$	diffusive mass flux of species k in axial/radial direction	$mol/(m^2s)$
k	reaction rate coefficient	−
k_H	Henry's constant	bar

k_r	coefficient for inertial resistances	$-$
k_W	heat transfer coefficient	$W/(m^2K)$
L	channel length	m
L_{pore}	pore length	m^{-6}
l_T	turbulent length scale	m
lhv	lower heating value	MJ/kg
$LIFT$	intake valve lift	mm
M	molar mass	kg/kmol
m	mass	kg
$m_{k,cum}$	cumulated mass profile of species k	kg
$m_{k,meas}$	measured mass of species k at time step t	kg
$m_{k,t}$	mass of species k at time step t	kg
$meas$	measurement profile	
mf_F	fuel mass fraction in the bulk of the oil film (F)	$-$
mf_G	fuel mass fraction in the gas mixture (G)	$-$
mf_L	fuel mass fraction at the interface of the oil layer (L)	$-$
n	chemical amount	$-$
N_G	number of gas species	$-$
N_H	Henry Number	$-$
N_R	number of gas phase reactions	$-$
OSC	oxygen storage capacity	gO_2/l
P	power	kW
p	pressure	bar
pme	pressure mean effective	bar
$q_{a/r}$	heat flux in axial/radial direction	$W/(m^2kg)$
Q_b	heat energy of the combustion	W/m^2
Q_{wall}	heat energy through the cylinder walls	W/m^2
R	specific gas constant	$J/(kgK)$

r	radius	m
r_{conv_k}	conversion rate of species k	%
$ratio_{gear}$	transmission ratio	–
Re	Reynolds number	–
RF	relative filling	%
rpm	engine speed	min^{-1}
s_L	laminar burning velocity	m/s
s_T	turbulent burning velocity	m/s
s_{pist}	piston position	m
s_{SP}	spark plug position	m
s_{stroke}	stroke	m
Sc	Schmidt-number	–
sim	simulation profile	
SM	safety margin	–
T	temperature	°C
t	time	s
T_{air_int}	intake air temperature in front of the throttle	°C
$T_{coolant}$	engine coolant temperature	°C
$T_{eng,warm}$	engine temperature at warmed-up condition	°C
T_{eng}	engine temperature	°C
T_{oil}	engine oil temperature	°C
T_W	channel wall temperature	°C
T_w	cylinder wall temperature	°C
TH	throttle position	%
TQ, tq	torque	Nm
U	internal energy	J
u	flow velocity	m/s
u_e	velocity of the flame front	m/s

V	volume	m^3
v	vehicle speed	km/h
v'	turbulent velocity fluctuation	m/s
v_act	actual vehicle speed	km/h
v_target	target speed of the driving cycle	km/h
v_1	flow velocity	m/s
v_a	axial velocity	m/s
v_r	radial velocity	m/s
v_{veh}	vehicle speed	km/h
WG	wastegate position	mm
X	mass fraction	—
x	axial coordinate	m
y_{bd}	ratio of burn duration	—
Y_{EGR}	residual gas content	%
$Y_{k,meas}$	measured mass fraction of species k	—
Y_k	mass fraction of species k	—
z_a	cylindrical coordinate in axial direction	m
z_r	cylindrical coordinate in radial direction	m

SUPERSCRIPTS

Symbol	Meaning	Unit
α, β	exponents laminar burning rate	
a, b	Arrhenius exponents for concentration	
cyl	cylinder	
$load$	precious metal loading	

SUBSCRIPTS

Symbol	Meaning	Unit
0	reference state	
1	actual	

acc	acceleration
act	actual
adj	adjusted
$aero$	aerodynamic
b	burned
c	charge path
cr	crevice
cyl	cylinder
e	entrained
eff	effective
eng	engine
exh	exhaust
F	flame
g	gear number
i	indicated
in	into
int	intake
$int-r$	intake restricted
$int-ur$	intake unrestricted
j	second
k	species
man	manifold
max	maximum
$meas$	measurement
min	minimum
$norm$	normed
opt	optimized
out	output

r	reaction index
rel	relative
req	requested
$roll$	rolling resistance
S	sphere
s	spark path
SP	spark plug
t	target
$tire\,i$	tire number
tot	total
u	unburned
veh	vehicle
W	channel wall
x	longitudinal direction
z	index of cylinder wall zone

1

INTRODUCTION

This work starts with the motivation for the presented simulation methodology (1.1) followed by the introduction of state of the art technologies for combustion engines in automobile applications in section 1.2 and the explanation of emission evaluation (section 1.3). In the last section 1.4, actual calibration processes are depicted and connections to virtual methods are shown.

1.1 MOTIVATION

Raising environmental awareness and finite nature of fossil fuel supply resulting in increasing regulatory requirements for vehicle emissions force the automobile industry to achieve emission improvements. Actual challenges follow from prospective new legal provisions for driving cycle measurements and emission recording (Worldwide harmonized Light-Duty vehicles Test Procedure (WLTP), Real Driving Emissions (RDE)) in the future [177]. However, reducing emissions of automotive vehicles requires new technologies for combustion engines. In state of the art spark ignition (SI) engines, advantages of turbochargers, fully-variable valvetrains and direct injection (DI) are combined to accomplish the best compromise between driving dynamics and emissions. As a consequence from these technological enhancements, the number of actuating variables of combustion engines is expanded resulting in more and more complex engine control systems. Combined with short development loops, virtual methods become increasingly important. In order to compensate high costs of hardware-based calibration work, a simulation environment for optimal parametrization of the engine management system is required. With its help, the system understanding can be supplied and new concepts can be studied.

The aim of this work is the development of a simulation methodology to analyze fuel consumption and emissions formation during driving cycles based on a detailed modeling of the combustion engine and considering calibration interactions. Normally, simple empirical combustion engine models based on extensive test bench studies are utilized in virtual calibration tools. On the other hand, predictive combustion models for homogeneous stoichiometric SI combustion are investigated since many years (e.g. [7, 18, 144, 161]) for the purpose of their application within a thermodynamic 1D gas exchange simulation. According to the literature [16, 17, 66, 109], it is also common to analyze emissions virtually. However, these models are focused mostly on steady-state operation and executed independently to examine individual effects.

The new approach of the simulation methodology presented here is to combine a calibration environment with a thermodynamic engine model including a predictive combustion sub-model and appending a virtual vehicle powertrain to simulate various driving cycles [24]. The coupled system is extended by calculations of engine-out emissions considering the formation of carbon monoxide (CO), nitrogen oxide (NO_x), and hydrocarbons (HC). Hereby, the seamless coupling of the different sub-models, as well as the extension of the combustion model towards dynamic operation is the main challenge. In the context of fuel consumption analysis, this virtual approach offers the possibility to gain insights in the transient thermodynamics engine behavior. Concurrently, the calibration of the engine can be adapted, optimized, and analyzed in this simulation environment. In an additional simulation, tailpipe emissions are determined to obtain a completely virtual evaluation of the global vehicle system. The purpose of this comprehensive methodology is to provide a physical based calculation and to represent a good compromise between computing time and quality of results. It is not demanded to achieve real time capability nor to accomplish a detailed modeling compared to 3D Computational Fluid Dynamics (CFD).

According to the following structure, the contents of this simulation methodology are described, details of the system are explained, and its quality and applicability are evaluated:

1. Introduction

2. Simulation Methodology

3. Modeling

4. Transient Simulation

5. Application Examples

6. Summary

First, a basic overview about relevant technologies of a modern combustion engine, the evaluation of emissions, and actual calibration processes are introduced (chapter 1). In chapter 2, the simulation methodology and modeling background are depicted. Thereby, requirements of each component are defined and different modeling approaches from the literature are discussed. The modeling of the sub-models and the coupling are presented in chapter 3. The calibration of the combustion module and the prediction of in-cylinder emissions are demonstrated by results of steady-state engine operation. In chapter 4, the transient performance in driving cycles is validated and matched to measurements. The practical application of this simulation methodology is shown based on further investigations (chapter 5). At the end, this work concludes with a short summary (chapter 6).

1.2 STATE OF THE ART - TECHNOLOGIES

Actual developments of internal combustion engines have to challenge with different aims competing against each other in parts. Figure 1.1 illustrates the four main targets (comfort, driving dynamics, fuel consumption, and emissions) for modern passenger vehicles, wherefore the powertrain has to achieve a compromise. However, these goals often generate conflicts regarding environmental, economical, and legal boundary conditions. As motivation for this work, one significant focus is on the improvement of efficiency, trying to reduce emissions without losing power (Golloch [44]). This is affected by technologies of the exhaust aftertreatment and by the combustion process which in turn depends on the parameters for the fuel-mixture generation. In the last years, a lot of measures were developed to increase the fuel efficiency, such as direct injection, supercharging or variable valve timing to control the gas exchange. An overview of the primarily important technologies considered in the presented methodology is provided in the following.

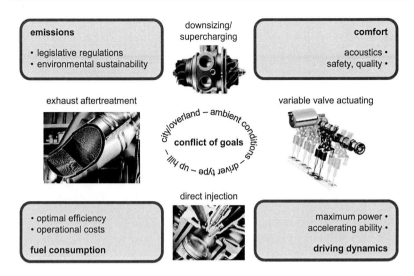

Figure 1.1.: Goal conflict in the development of combustion engines [36]

Supercharging

Supercharging is one of the proposals to reduce fuel consumption while keeping the power at least on the same level. This is achieved by scaling down the displacement in combination with increasing the cylinder pressure. On the one hand, it enables to minimize losses of gas exchange, friction and wall temperature resulting of smaller cylinder volumes. Thereby, the engine performance is reduced without countermeasures. Consequently, the indicated mean effective pressure pme has to be raised. It is proportional to the air density in the cylinder $\rho_{air,cyl}$ and, using the ideal gas equation, as well as to the pressure before compression:

$$pme \sim \rho_{air,cyl} = \frac{p_{air,cyl}}{R \cdot T_{air,cyl}} \tag{1.1}$$

Supercharging enables now to boost the cylinder pressure and accordingly air density resulting in higher mean pressures and engine performance. In equation 1.1, a temperature dependency of the mean effective pressure based on the air temperature in the cylinder can be detected. Hence, a cooling of the compressed air is required to obtain the power gain ([44, 60, 75, 170, 173]).

In automobile vehicles, several concepts exist for supercharging. Their classification can be realized according to gas dynamic effects (resonance supercharging), mechanical turbocharging (compressor), exhaust-gas turbocharging or a combined system (turbo compound system) (Hiereth and Prenninger [60]). In this work, only engines with an exhaust-gas turbocharging are considered (Klauer et al. [79]). This technology is based on a thermodynamic link of the combustion engine and a turbo-machine taking advantage of the kinetic energy in the exhaust gas for the pressure charging. On the one side of a shaft, a turbine is powered by the exhaust flow of the combustion process. At the other end of this mechanical link, a compressor wheel is transforming the rotating energy into a higher pressure load of the fresh air on the intake

3

side. The compressed air is then cooled down by an air intercooler before it passes the throttle and flows into the intake manifold. The kinetic energy can be converted more effective, when using a twin scroll turbo charger. Due to the dividing of the exhaust gas flow into two air-streams, interacting effects between the cylinders exhaust strokes to the gas exchange can be prevented [29, 79, 80]. For the purpose of controlling the power output of the turbine, a bypass called wastegate is integrated.

Full variable valve actuating/ gas exchange

During the gas exchange of a combustion engine, a enhanced fuel-mixture generation can be achieved by an directed charge motion. Thus, the gas exchange is an important parameter for a clean and efficient burning of fuel. It is effected significantly by the opening time of intake and exhaust valve. These can differ for different engine operation. For this reason, actual engines are often equipped with a full variable valve actuating (VVT) (Klauer et al. [81], Steinparzer et al. [159]). It controls the opening phase by adjusting the valve timing (specified by the difference between the maximum valve lift and gas exchange top dead center (TDC) in crank angle at intake and exhaust valve). Additionally, maximum lift of the intake valve can be changed. This adjustability is illustrated in figure 1.2 and requires three parameters called *IVT/EVT* for the valve timing (intake/exhaust) and *LIFT* for the intake valve lift, which have to be set by the ECU.

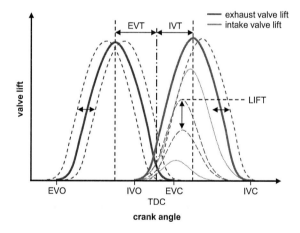

Figure 1.2.: Actuating of a full variable valve train according to a VVT of the BMW Group [81]

Valve-opening overlap (when both, intake and exhaust valve, are open at the same time) is important for the control of the content of residual gas by internal exhaust gas recirculation. This variability enables a compromise of fuel consumption and dynamic aspects, particular at low engine speeds and loads. At part load, the concentration of residual gas gets higher with a large overlap. The afterburning of the exhaust gas reduces CO_2 emissions, but also the power output of the engine.

A special effect can be seen for turbocharged engines with large overlaps at low engine speeds and high loads. Due to a positive pressure gradient between intake and exhaust manifold, the cylinder gets flushed by fresh air. This is called scavenging. When nearly no content of residual gas is left, the combustion chamber can be filled by more fresh air and the power output rises. Also the air flow increases by scavenging leading to a faster speeding-up of the turbo charger. Additionally, the enthalpy before the turbine gets bigger by post-oxidation of unburned fuel and the excess of air resulting in higher boost pressures. Consequently, scavenging improves the response behavior of vehicles with turbo charged engines. This effect is getting smaller with increasing engine speeds, since the time slot of the positive pressure drop is shortened. An optimization can be achieved by dividing the exhaust gas flow as for the twin scroll turbo charger mentioned above [10, 100, 134, 148, 149]. However, scavenging has also a negative influence on emissions and degradation of the catalyst. Because of the low content of residual gas, post-oxidation is displaced from the cylinder to the exhaust system. The raised exhaust gas temperature forwards the generation of nitrogen oxides and stresses the catalyst thermally [27], Brandt et al. [10].

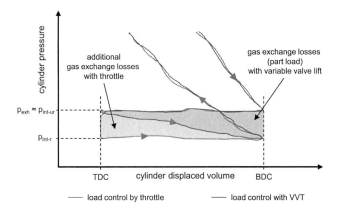

Figure 1.3.: Gas exchange losses in the part load according to [46, 109, 170]

Adjusting the lift of the intake valve can produce two positive effects. First, a throttle-free load control can be realized to reduce losses of gas exchange. But charge motion can be improved as well, especially at small valve lifts. There, swirl can be produced in the combustion chamber by phasing (different lifts for both intake valves). Combined with tumble, it generates a higher turbulence for better residual gas compatibility and homogeneity of fuel-mixture. Different systems can be found to realize the variable valve lift. The engine investigated in this work is equipped with mechanic actuators consisting of a servomotor, a roller cam follower with lever, an eccentric shaft and a return spring. The actuating is controlled electric by the parameter $LIFT$ from the ECU. Due to the adjustable valve lift, the opening time of the intake valve can be adapted with regard to the required mass of fresh air in the cylinder. Thus, the pressure in the intake manifold has not to be restricted [78, 154, 169, 170]. This is different to an engine with load control by a throttle placed in front of the intake manifold. Here, the throttle controls the mass of fresh air trapped in the cylinder as a function of pressure and air flow, accordingly. Losses of gas exchange depend on the pressure difference of intake and exhaust manifold. It increases with smaller angles of the throttle in partial load [46, 109, 170]. In figure 1.3, an exemplary gas exchange loop of a spark-ignited engine with full variable valve actuating is shown. When the intake valves are open, fresh air with the unrestricted pressure of the intake manifold is flowing into the cylinder throttled by the valve orifice area. Compared with the area of the load control with throttle in the p-V-diagram, the losses of the gas exchange are reduced significantly (Göschel [46], v. Basshuysen and Schäfer [170]).

Technically, engines with a full variable actuation of the intake valve could be lacking of a throttle. Because of safety reasons in case of malfunction of the VVT and dynamic matters, a throttle is still installed and has to be controlled (Kfz-Technik [77]). State of the art, the actuation is done electrical by the parameter $THROTTLE$ from the ECU.

Ignition

Contrary to diesel engines, internal combustion engines powered by gasoline need an externally-supplied ignition. Near the end of the compression stroke, the air-fuel mixture is ignited by the spark plug in the combustion chamber due to carefully timed high-voltage between two electrodes. The generated spark produces high temperature plasma responsible for the start of an exothermal chemical reaction of the gas composition. Thereafter, a self-preserving flame front is diffusing in the combustion chamber (v. Basshuysen and Schäfer [170]).

Usually, the ignition timing describes the moment of the spark referred as a certain crank angle relative to piston position at TDC (positive = before). It has a significant influence on the combustion process and consequently, on the performance of the engine. With an earlier ignition, the maximum pressure in the cylinder is rising caused by a previous center of combustion mass. Best case of fuel consumption is a center of energy-conversion mass at a position of approximately eight degrees after firing TDC (v. Basshuysen and Schäfer [170]).

Furthermore, several driving situations require variable ignition timing. E.g., an earlier ignition timing is useful for a powerful output. Otherwise, high exhaust gas temperatures generated by late ignitions after TDC are essential for catalyst heating. On both sides, early and late, the range of adjustment is limited. Early combustion is restricted by knocking, for example. After ignition by the spark, the flame front is diffusing into the combustion chamber. Because of increasing pressure and temperature in the cylinder, an autoignition of the unburned air/fuel mixture can occur outside the envelope of the normal combustion front. As a result, shock waves create the characteristic metallic sound and cylinder pressure increases dramatically. These knocking effects can be completely destructive for the engine.

On the other hand, late ignition timing is limited by the complete combustion of the air/fuel mixture and temperature of exhaust gas (Merker et al. [109], v. Basshuysen and Schäfer [170]). Based on late spark timings, the center of energy-conversion mass is delayed to the expansion phase of the working cycle. Less energy of the combustion is transferred to mechanical work (engine torque drops), but the enthalpy of the exhaust gas raises and consequently, its temperature. This effect increases with growing engine load and speed and is realized in terms of catalyst heating and speed up of the turbocharger. In addition, extreme late or early ignition timing can result in misfiring, especially at low engine speeds and loads. That means, the air/fuel mixture does not burn (Miklautschitsch [111]).

Direct fuel injection

Currently, it is common usage in passenger vehicles to inject the fuel directly into the cylinder. This generates several advantages like the decrease of the temperature in the cylinder by the evaporation enthalpy. Hence, compression ratio can be increased and efficiency can be improved over the whole engine map (Golloch [44]). Another benefit of direct injection is the enabling of variable operating modes. For instance, stratified charge can reduce fuel consumption under part load conditions, but is challenging due to emissions. The engine investigated in this thesis is operated homogenous and stoichiometric with single injection in most cases. Only at high loads and engine speeds, fuel enrichment can be required by reason of heat protection. In some special operating modes like catalyst heating, a second or even third injection event during each working cycle can be added. In combination with turbo charging, direct injection enables scavenging (see section 1.2). Due to last injections after closing of the exhaust valve, unburned fuel can be avoided in the air flow while flushing the cylinder.

It exists a couple of different basic design concepts of direct injection that can mainly be differentiated into air-formed, wall-formed and spray-guided processes. The first two ones can be challenging due to wall wetting by fuel drops. Strong oil dilution and high hydrocarbon emissions can be the outcome. In this work, a spray-guided system is implemented in the considered engine and the injector is positioned central near to the spark plug.

1.3 STATE OF THE ART - EMISSIONS

Vehicle emissions

In internal combustion engines, fuel is mixed with air to get burned. Gasoline used in vehicles consists of a composition of many different chains of hydrocarbons. If hydrocarbon molecules are burned completely, only carbon dioxide and water are formed theoretically. Due to not ideal conditions, following main components are emitted from SI combustion:

- carbon monoxide (CO)

- unburned hydrocarbons (HC)

- nitrogen oxide (NO_x)

Furthermore, the emissions include nitrogen from the air and oxygen in case of engine operation with excess of air (v. Basshuysen and Schäfer [170]). The simplified reaction mechanisms of the combustion process in the cylinder and the resulting exhaust gas composition in volume percent (vol.%) at a stoichiometric mixture of a SI engine are shown in figure 1.4. Despite all improvements in an efficient combustion, emissions that are harmful to the environment will remain. In figure 1.5, the interconnection of the main pollutants NO_x, CO, and HC and the air-fuel equivalence ratio with assumption of a homogenous mixture is exemplified for a SI engine at steady-state operation. At nearly stoichiometric mixture, the emissions of CO and HC can be minimized, while the production of NO_x reaches its maximum. In order to reduce the toxic pollutants of vehicles, an exhaust aftertreatment is required. Nowadays, a common way is the application of a three-way catalytic converter (TWC) at internal combustion engines fueled by either gasoline or diesel. This TWC works most efficient, when its body is heated up and the combustion engine operates within a narrow range of a stoichiometric

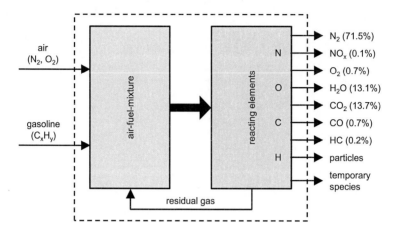

Figure 1.4.: Typical exhaust gas composition of the SI combustion process at $\lambda = 1.0$ in vol.% (v. Basshuysen and Schäfer [170])

air/fuel mixture. Even a slight oscillation outside of this narrow margin provokes a rapidly falling of the conversion efficiency. This effect is controlled by closed-loop systems for the fuel injection in the ECU using one or more oxygen sensors in the exhaust gas. Under ideal conditions and a heated TWC, convection rates near 100% can be achieved.

In purpose of complying the todays emission regulations, the period of cold run has been shortened by following measures:

- A location of the TWC near the engines exhaust manifold.

- Air injection in front of the converter heating-up the TWC faster by the exothermal reaction with unburned fuel.

- Double-walled exhaust tubes to prevent cooling of the hot exhaust gas.

- Electrical heated catalyst converter

Vehicles with diesel or lean burn SI engines require additional NO_x adsorber or a selective catalytic reduction (SCR) due to the high concentration of nitrogen in the exhaust gas. Since catalytic converters cannot clean up elemental carbon, particular filter against soot are appended. In this work, an engine with one TWC located near the turbocharger at the exhaust manifold is investigated.

Government regulations for emissions

California was the first state trying to reduce effects of the growth in motorization to the air quality. Corresponding to increasing environmental and health awareness, exhaust-gas limits for vehicles were established starting in USA in the year 1961 and followed by other states like Japan in 1966 and Europe since 1970 (v. Basshuysen and Schäfer [170]). Since 1974,

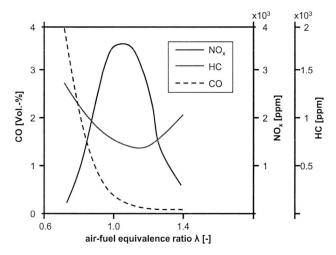

Figure 1.5.: Engine-out emissions dependent on air-fuel equivalence ratio (Merker et al. [109])

catalyst converters are prescribed by government in some parts of the USA. In Germany, it is regulated by law that all new cars after 1989 have installed one and only vehicles with TWC are admitted since 1993. In Figure 1.6, the continuous intensification of emission regulation for SI engines in Europe in the last years based on the restricted amounts of the EU3 is presented. While the gaseous pollutants have to be reduced by half, new limits are added for particulate matter (PM) in EU5 and particle number (PN) in EU6.

Emission limitations of vehicles are assessed by driving cycles which are also designed to specify fuel economy. The advantage of these driving cycles is the fixed test procedure. A theoretical distance is represented by providing a vehicle speed profile versus time. The cycles have to be driven on a flat road without influences of weather (wind, outside temperature) or traffic conditions. In order to provide constant ambient conditions and a repeatability of each measurement, the cycles are performed on chassis dynamometers. An electrical machine emulates resistance due to the vehicle characteristics like e.g. aerodynamic drag and inertia. As a function of the current speed, a fan is blowing in front of the vehicle supplying an airflow to the intakes of the combustion engine and radiators. Tailpipe emissions are collected and evaluated to supervise the regulations and to determine fuel consumption of the vehicle.

Several driving cycles with corresponding exhaust-gas limits are designed by different countries and organizations in the world. They can be classified into two types: Transient driving cycles with many changes in speed, representing typical on-road driving. However, modal driving cycles include protracted periods at constant speeds after accelerations. Last one are derived theoretically (for example the European NEDC), whereas transient driving cycles are often direct measurements of a representative driving situation like the American FTP-75. Other important modal driving cycles are the Japanese 10-15 Mode and JC08. Automobile manufactures selling their vehicles in the whole world have to consider all government regulations. In best case, they can be observed all together only by variation of engine calibration without appending additional parts for exhaust aftertreatment.

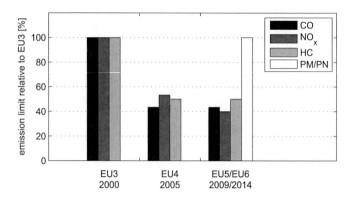

Figure 1.6.: Limitation changes from EU3 to EU6 for SI engines referred to the EU3 amounts [170]

In this work, the NEDC and more dynamic driving cycle anticipating future emission regulation are considered for transient investigations:

New European Driving Cycle (NEDC): Actually, the NEDC (New European Driving Cycle) is still the official driving cycle in Europe to assess emission levels and fuel economy in passenger vehicles (Europäisches Parlament [33]). The NEDC is theoretically derived from the typical driving profile in Europe. It represents an urban driving part (UDC), repeating the ECE-15 four times, and one Extra-Urban driving period (EUDC). The profile of vehicle speed versus time is shown in figure 1.7. The total distance is theoretical about 11km and the total cycle lasts 1180s. If the vehicle is equipped with a manual gearbox, gears and time periods for shifting and disengagement of the clutch are prescribed in the test procedure. It also defines a start temperature of the vehicle at 20-30°C to consider efficiency effects of various parts (e.g. engine, transmission, catalyst converter, etc.) due to cold conditions. During the test all ancillary loads like air conditioning, lights, and entertainment electronics, for example, are turned off (Europäisches Parlament [33]).

Random Cycle Aggressive (RCA): Emissions from vehicles varies depending on the driving profile, driving characteristics, ambient conditions, and so forth. In the last years, investigations show an exceeding of the regulatory emissions standards during real-life driving situations. These are insufficiently represented by the NEDC with its long constant speed cruises, many idling events, and only low engine loads. Currently, new approaches to measure the wide range of potential Real Driving Emissions (RDE) are in development by a EU working group (Weiss et al. [177]). One solution is an On-Road based test procedure applying a Portable Emission Measurement System (PEMS) (Weiss et al. [176]). Hereby, the investigated vehicle is driven on public roads representing "real conditions", while measuring the tailpipe emissions. Due to unsteady environment situations, a laboratory based test with repeatable and reproducible conditions is also discussed. Therefore, a random driving cycle has to be generated. Actual proposals to achieve different severity consist of various vehicle speed profiles representing soft, normal or aggressive driving behavior, and differing climatic

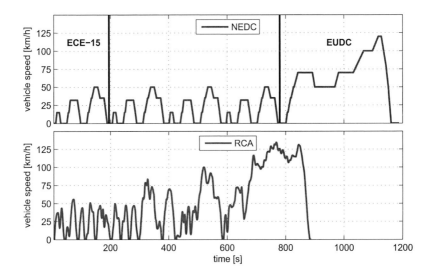

Figure 1.7.: Vehicle speed profiles of the NEDC and RCA

conditions of temperature and humidity. In this work, a random cycle aggressive (shown in figure 1.7) started at ambient conditions is performed as counterpart of the NEDC. The implementation of the RDE procedure for new vehicle approvals is expected not before 2017.

Measuring of emissions

The regulated components of the exhaust gas have to be measured by separated analyzers. At the end of the tail pipe, the exhaust gas is extracted by a heated hose to avoid condensation of hydrocarbons. After diverting a branch current to analyze amounts of hydrocarbons and particulates, the exhaust gas is cooled down and the condensate is collected by a water separator.

Measuring of carbon monoxide (CO), **carbon dioxide** (CO_2) **and methane** (CH_4): The nondispersive infrared detector (NDIR) is based on the substance dependency on radiation absorption of gases. The usual structure of a NDIR analyzer is shown in figure 1.8. Infrared radiation is rayed through two unconnected chambers, one filled with the exhaust gas to be analyzed and the other one contains an inert gas (e.g. nitrogen, oxygen, and hydrogen) for comparison. The inert gas must not absorb radiation in the interested spectral range. Behind this filtering, the light is detected in two further chambers separated by a membrane. Both chambers are filled with the exhaust component which has to be measured. Because of the unequal intensity of the infrared radiation, the two chambers are heated unbalanced generating a pressure gradient. Finally, the absorbancy can be measured by the deformation of the membrane between the two chambers. This principle of the NDIR-adsorption allows the measuring of several exhaust-gas components at the same time by arranging multiple chambers in a circle (Klingenberg [82]).

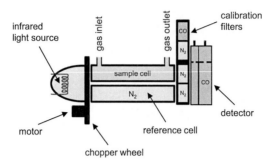

Figure 1.8.: Schematic structure of a nondispersive infrared detector (NDIR)

Measuring of hydrocarbons: The amount of Hydrocarbons in the exhaust-gas can be analyzed by a flame ionization detector (FID). Its construction of a burner and an electric circuit is illustrated in figure 1.9. The exhaust gas is burned in a diffusion flame of hydrogen

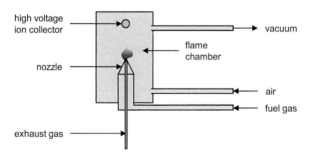

Figure 1.9.: Schematic structure of a flame ionization detector (FID)

(fuel gas) between two electrodes with voltage application. The content of hydrocarbons gets ionized by the flame resulting in a charge transfer to the cathode. The measured current is proportional to the concentration of hydrocarbons in the exhaust gas. This is a very robust process due to the self-cleaning by the combustion and a constant feeding of combustion and exhaust gas. However, only the concentration of the total amount of all hydrocarbons contained in the exhaust gas is detected. The determination of individual HC compounds is not possible (Klingenberg [82]).

Measuring of nitrogen monoxide (NO) and nitrogen oxide (NO_x): The chemiluminescence detector (CLD) is using the effect, when molecules emit energy in the form of electromagnetic radiation by chemical reactions. The reaction of nitrogen oxide with ozone forms oxygen and nitrogen dioxide which is partially in an exited state. The excess of energy emit the molecules as fluorescent radiation proportional to the concentration of nitrogen monoxide. This effect is measurable optically in the CLD. This chemical reaction works only for NO molecules and consequently, the NO_2 content in the exhaust gas has to be reduced to NO by a thermal converter at first (Klingenberg [82]).

Measuring of oxygen (O_2): The oxygen content in the exhaust gas can be measured with the help of its paramagnetic characteristics. In an inhomogeneous field of a magnet, paramagnetic compounds are attracted to the pole, diamagnetic ones are rejected. All technical gases have diamagnetic characteristics except of oxygen. The magnetic separation of oxygen can be measured by a hot-wire anemometer or a pressure differential analyzer.

Calculation of transient emissions

Usually, simulations to detect fuel consumption and emissions in driving cycles use quite detailed models of the vehicle powertrain, from a mechanical point of view. However, thermodynamical aspects of internal combustion engines are neglected and modeled on a basically way. In general, measured maps of an engine representing fuel consumption, emissions, and performance are implemented to simulate the power unit [13, 15]. In cases of a newly developed engine that is not available in hardware yet, these maps have to be estimated or transferred by data of the forerunner series. In the automobile industries, several approaches in combination with different simulation tools are applied [35, 54, 56, 135, 146, 147, 171, 172, 175].

The implementation of look-up tables has some obvious weaknesses for the estimation of transient vehicle pollutants, since these maps consist of steady-state measured operation points. Dynamical effects like fast load changes are not considered. For example, each individual operation point is measured with exactly one combination of actuator parameters for the engine under standard conditions. In practice and accordingly in a driving cycle, different environment conditions can appear compared to the engine test bench or engine operating points can be located between the measured ones. Due to various input, the actuator parameters from the ECU can slightly fluctuate. Compared to the map from the engine dynamometer, the combustion can differ and can produce an unequal amount of emissions or power output. Definitely, this is important in cases of special operating modes such as catalyst heating, when late ignition timings are actuated. Another weakness is the measurement of maps with warmed-up engines. Thus, the engine oil, cooling fluid, and the engine body are at normal running temperature. In driving cycles, a start with a cold vehicle conditioned at room ambient temperature of approximately 25°C is regulated by law. Hence, increased friction and heat loss at cold engine conditions can lead to more fuel consumption and emissions.

1.4 STATE OF THE ART - CONTROL

The actuation of the combustion engine is controlled by an electric engine control unit (ECU). The functions of the control have to be adapted to the corresponding engine during a calibration process.

Engine control unit

The ECU has to manage the different requests to the engine from driver, emission regulations, driving dynamic control, and emergency systems. At the beginning of combustion engines, the control is realized by mechanical, pneumatic, and hydraulic adjusting. Based on the raised engine actuator parameters nowadays, the physical phenomena of gas exchange and combustion have to be described with the help of mathematical and computer engineered functions in the ECU. Not only the performance and complexity of autonomous functions for the actuation is grown continuously in the last centuries, also the combination and connection of these different functions are advanced and integrated in large modules (Gerhardt et al. [43], Robert Bosch GmbH [139]).

The software system is implemented in a computer with an aluminum casting box. In the body, a cooling structure is integrated for heat transfer to the surrounding air preventing overheating. It also contains plug-in positions for input and output connections. Required parameters to determine actual operation conditions like e.g. engine speed, temperatures, and pressures are measured by sensors and connected to the ECU by wires (Isermann [69], Roithmeier [140]). Based on a defined signal processing, input parameters are converted and the target parameters for the actuation of the engine are transferred to the particular output slots after calculation. In purpose of safety issues, each incoming signal and output variable is monitored and qualified permanently (Roithmeier [140]). Due to legal restraints, a so called on-board diagnostic (OBD) slot for readout or coding of required functions in the ECU is implemented (Zimmermann and Schmidgall [183]).

The aim of the ECU is to determine engine and vehicle conditions by sensors or calculations and to control them in line with demand. Usually, this is realized by linear controllers (e.g. PID controller). In purpose of considering nonlinear properties, parameters have to be generated for several engine operating points and stored in maps. These variables can be defined as functions of e.g. engine speed and load, cylinder charge, and engine temperature. Partitioned in different modules, it results in a complex system with many adaptable calculations and adjustable parameters.

Engine calibration process

The functions of the ECU are parametrized by many look-up tables and constant parameters. These can differ from one engine version to another and have to be adjusted for each vehicle type as well. In the last years, this calibration process is getting more and more complex due to the rising number of variable engine actuators and growing range of vehicle derivatives. Even though engine characteristics is limited by its construction and technical configuration, the calibration of the single functions in the ECU can still affect it considerably. The challenge of the calibration process is to achieve a good compromise of several constraints like regulations by law, demands for driving performance versus fuel consumption and standards in comfort and acoustics (compare figure 1.1). A classic conflict of these aims is to optimize dynamic behavior of the engine in terms of power output without increasing fuel consumption (ETAS [32]).

The calibration process is supported by different practices according to its task and stage of development. They can be differentiated by the used resources like engine (stand-alone), vehicle or virtual methods (Roithmeier [140]):

Engine test bench/engine dynamometer: The hardware consists of the entire engine including all its attached parts like intake/exhaust system and cooler. It is controlled by the original ECU, while selected engine management parameters can be varied and results are recorded automatically by special calibration software. The engine is linked mechanical to an electric generator to simulate a variable load. Power and torque are measured directly from the engines crankshaft by the engine dynamometer. According to its request, specific measurement instrumentation can be added. The test bench is for stationary operation and conduces to fundamental analysis and calibration of basic parameters. In purpose of complete engine operating maps, the measurement process is often fully automated.

High dynamic engine test bench: Usually, transient operation of the engine is tested and measured in vehicles. Due to cost effectiveness and complexity of vehicle prototypes, high dynamic engine test benches are applied to operate the engine transient. Even complete driving cycles can be performed by simulating the remaining drive train of the vehicle. It is implemented by a high dynamic electric motor controlled by a real-time simulation of the driver and vehicle. The upgrading of the classic engine test bench is useful for the repeatability of transient test cycles under defined conditions.

Vehicle: Despite high dynamic engine test benches, calibration work will always be performed in the vehicle. Therefore, measurements are done in defined maneuver on test tracks or general driving situations are tested in the public road traffic. For this purpose, special test ECU's prepared for changing calibration parameters/software are installed in the vehicles. Vehicle tests provide the evaluation of demands for driving performance, comfort, and acoustic issues.

Exhaust chassis dynamometer: Regulations by law define driving cycles to control emissions and to determine fuel consumption. These measurements are conducted on an exhaust roller dynamometer to ensure repeatability and comparability. The vehicle is parked on the roller measuring the power delivered to it by the drive wheels. The dynamometer is simulating air and roll resistance of the vehicle. Emissions sampling equipment is integrated and fuel consumption can be analyzed.

Hardware-in-the-Loop (HiL): This test bench is mainly for functional checks of the engine management system. It includes the ECU in hardware integrated in a closed-loop simulation of its sensors and actors. On demand, additional hardware like engine actuators (e.g. throttle) or vehicle cockpit is attached. Driving cycles can be analyzed as well, if virtual models of engine and vehicle are added to the simulation environment.

Virtual calibration process: In order to reduce test costs, the calibration work is transferred entirely in a virtual environment. Each component, from engine and actuators to the ECU and even the whole vehicle, are modeled and implemented in simulation tools. The calibration process can consist of a simple variation of parameters for engine actuation (engine stand-alone simulation) or include the changing of complete maps and coefficients in the ECU (simulating the coupled control functionality). In figure 1.10, different tools of the virtual calibration process are illustrated. It supports the conventional calibration with the fast analysis of measurements from the test bench by various techniques of calculation, conversion, and filtering. Additional data can be provided by simulations of the investigated system with variable modeling depth. According to the requested level of detail, the models can be adapted by independent sub-modules for e.g. the combustion process, friction losses, engine control, and vehicle. This enables the analysis of complicated issues and improves the understanding for interaction and coordination of the complex system. Overall, a high grade of automation is aimed in this virtual process including evaluations, validations, and optimizations (Kratzsch et al. [87]).

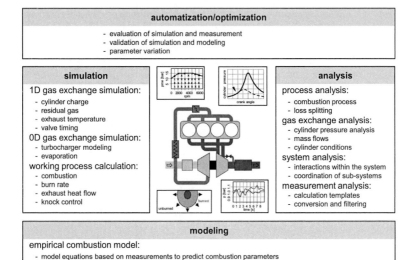

Figure 1.10.: Virtual calibration process [87]

Often, a detailed engine model is applied in combination with a simple engine control. According to the complexity of the simulation environment, real-time capability cannot be achieved in most cases. Simulations can be performed for e.g. calculations of cylinder charge and torque to supply the calibration of gas exchange parameter (valve timing), combustion process (knock detection) and exhaust aftertreatment (gas temperature). Compared to the potential variations of analyses, the engine model can be matched by only a few operating points from a measurement to predict the entire engine map. Based on the physical modeling, the simulation environment can be applied in follow-up projects or in the evaluation of future concepts. Due to the rising complexity to actuate modern engines (described in chapter 1.2), the application of the virtual calibration process will be spread continually in the future to improve the efficiency of powertrain development (Roithmeier [140]).

2

SIMULATION METHODOLOGY

The variability of actuating parameters for combustion engines is enlarged due to the technological enhancements introduced in chapter 1.2. Thus, the increased number of degrees of freedom in the system results in a high complex modeling and controlling in the engine control unit. Combined with short development loops, virtual methods will play an important role for the development of vehicle powertrains. Consequently, a simulation environment is required to support the calibration work and compensate high costs of testing.

In this chapter, the simulation layout is described and requirements of each component are defined. First, the structure of the control system according to the architecture of a modern ECU and the necessity of a driver actuation is introduced (2.1). The physical system is explained by discussing different modeling depths of the combustion engine and specifying the main function of the vehicle powertrain model in section 2.2. The new approach of this work, combining a virtual calibration environment with a thermodynamic engine model, is presented in the section of simulation method and environment (2.3). Modeling background for the simulation of in-cylinder and tailpipe emissions is provided in section 2.4.

2.1 VIRTUAL CALIBRATION: MODEL OF THE ECU AND DRIVER ACTUATION

Virtual calibration methodologies can imply to implement a model of the ECU functionality. This can include the adopting of the original source code entirely or parts of it. Another approach applies the simulation of the ECU calculations by substitutional functions (Dorsch et al. [23]). In order to investigate driving cycles, the behavior of a driver has to be reproduced and its actuation has to be incorporated in the simulation environment.

2.1.1 *Engine Control Unit*

The function of the ECU is to translate the driver input into the desired vehicle reaction with respect to fuel consumption, emissions, comfort, and security aspects. This is accomplished by identifying physical effects and their influencing variables. The development process of ECU functions is illustrated in figure 2.1. It starts with the description of physical characteristics and functionality of a system, followed then by the definition of important parameters and modeling of the calculation, and ends with the coding. Usually, the modeling is physically-logical based and its extent depends on the required accuracy, available computation resources, and acceptable calibration costs. Finally, the functions are translated into the source code to implement the calculation into the ECU environment.

Nowadays, the functions in the ECU are based on a generic system platform with open architecture components (AC). Since the interfaces are defined explicitly, the exchange or expansion of individual modules is enabled. This aggregate concept allows automobile manu-

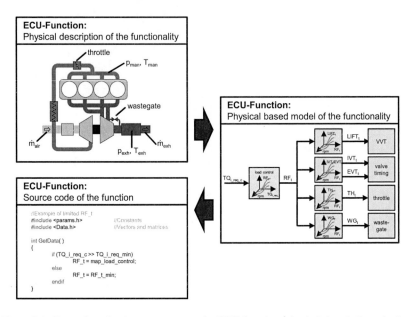

Figure 2.1.: Exemplary development process of a ECU function (physical description, physically based modeling, and source code of the functionality)[69]

factures to append their own functions and modeling (Isermann [69]). The individual modules consist of function packages, so called base components (BC), to describe the behavior and control the engine subsystem. This is pointed out for the air system in figure 2.2, exemplarily. In order to define the physical characteristics of the air flow in the engine, target values of the charge and further merged calculations to control the desired behavior of the subsystem (like "engine charge control" and "gas exchange valve control") are integrated into the function package "air system model". Each base component combines individual functions of the ECU, the "functional components" (FC). Here, all algorithms, variables, and parameters to achieve the functional task are implemented. For example, the mentioned base component "air system model" contains functions to calculate the gas condition in the manifold ("Air-Mod_IntMnf"), the fresh air flow into the cylinder ("Air-Mod_DetCylCh"), the rate of the exhaust gas recirculation ("Air-Mod_EGR"), and many more to determine the air system (Isermann [69]).

There exist two fundamental approaches to model ECU functions compared in figure 2.3 according to Isermann [69]. A conventional method is the description of the total system straight mathematically based on measurements without physical background. The main dependencies are defined for normal state (normal pressure, normal temperature, etc.) using only a few physical parameters. Each discrepancy from these conditions has to be corrected by adjustment maps. This results in a complex system with many adaptations and makes the emulation in the virtual environment difficult and expensive. Hence, a model based approach consisting of independent functions implemented into coupled modules is adopted for the ECU

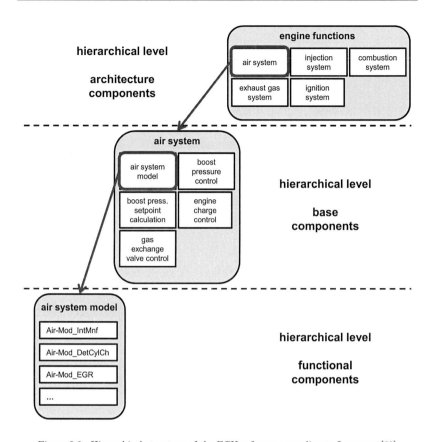

Figure 2.2.: Hierarchical structure of the ECU software according to Isermann [69]

model in this work. Thereby, the original system architecture is reproduced according to the hierarchical structure introduced above. This physical description into segmented subsystems enables the simplification of individual functions. Further advantage is provided by an easy exchange of sub-modules due to modeling requirements and their separated validation.

In the following, the concept of the organizational structure in the ECU model is described by introducing the main evaluation components important for the control of a combustion engine. Figure 2.4 illustrates the architecture level of a modern ECU distinguishing the main functions for a SI engine into separated modules. According to Isermann [69], the basis software includes hardware dependent packages involving monitoring, diagnostics, and communication. It connects the individual hardware components of the ECU and supplies the in- and outputs for the calculations. It imports the data of the sensors and exports the control values of the actuators. Their parametrization is usually provided by the manufacturer of the ECU. The calibration of the control unit to adapt it to an explicit engine or vehicle type is

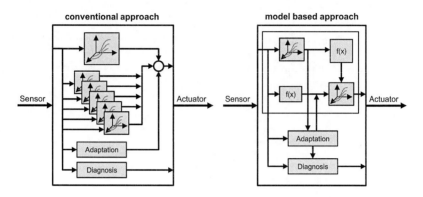

Figure 2.3.: Comparison of the conventional (map based) to the model based approach for the modeling of ECU functions [69]

done in the application software. It includes all functions to control the engine and handles the required vehicle information.

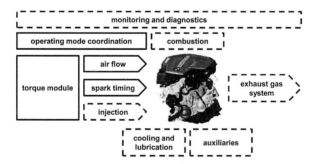

Figure 2.4.: Architecture of main ECU functions in SI engines according to Roithmeier [140] (solid lines: relevant, dashed lines: not relevant in this work)

As shown in figure 2.4, four main modules of the architecture components are relevant in this work:

- Operating mode coordination

- Torque module

- Air system

- Ignition system

In the operating mode coordination, special operation modes of the engine like e.g. catalyst heating, overrun fuel cut-off, and emergency program can be activated [140].

The central function in the computation of the ECU to control the operation of a combustion engine is the torque module. It is the connection of the driver input and other important parameters used in the ECU like cylinder charging, ignition timing, and fuel injection. Here, all calculations are predicated on the determination of the engine torque. This corresponds to the concept of a typical torque based architecture represented in figure 2.5. It was introduced

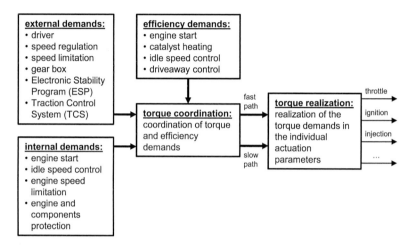

Figure 2.5.: Torque based function structure of the ECU [29, 68, 138, 139, 170]

in the 90's due to the development of the electronic accelerator pedal and is common in actual ECU's [29, 42, 43, 106, 138]. While former systems control the physical actuators directly by separated functions, the torque demands of different requests are coordinated and summarized to a target torque depending on their prioritization (Eichlseder et al. [29], v. Basshuysen and Schäfer [170]). In the torque module, the desired engine torque depending on external demands of e.g. driver, auxiliaries, and driving situation has to be evaluated with respect to internal requests like safety functions and operation conditions (e.g. engine temperature). In order to convert the efficiency demands controlled by the manager of the different operating modes, additional functions are implemented into the torque calculation. Before the actuation parameters are calculated, the target torque is splitted into two paths by the torque coordination, one for the air mass ("slow path") and the other one for the spark timing ("fast path"). The predicted torque of the torque module is finally realized at the end of this calculation path by the activation of the individual actuators (determined in the modules of air, ignition, and injection system).

As input for the computation, operating data from engine, vehicle and environment have to be sensed. When simulating the control behavior of an ECU, this information has to be provided, too. The number of these input parameters depends on the level of detail in the modeled functions. Therefore, the required interface specifications have to be determined and implemented as input and output connections. Since the combustion engine is also simulated in this work, additional interesting sensed properties (e.g. not or only difficult measurable data like residual gas content) can be provided compared with an engine test bench. This advantage should be considered in the modeling of the ECU modules and development of the interchange for the coupled system.

The primary requirements for the ECU model and resulting implications are summarized in table 2.1.

Requirement	Implication
transparent	clearly arranged with functional structure and content
modular	functions organized in hierarchical and exchangeable components
functional	simplified functions and modeling due to relevance
user-friendly	parameterization directly by the source calibration
accurate	reproducing results of real ECU

Table 2.1.: Summarizing of key aspects required for the ECU model.

2.1.2 *Driver actuation*

In order to simulate driving cycles, the vehicle has to follow a speed profile. In this work, a methodology is built to evaluate vehicle simulations nearly independent of measurements. Especially, the possibility of investigating several variations of driving cycles, engine application, and technical concept studies are of interest. Following matters prove the necessity of a control for driver actuation:

- When analyzing new driving cycles (like the upcoming regulation changes by WLTP and RDE), measurements of each vehicle derivation are often not available. Consequently, the driver actuation is unknown, too.

- Even if kinematics and resistance of the vehicle powertrain are constant, the driver actuation can differ due to various engine calibrations.

- Testing new technical concepts affecting the engine can result in different driving actuation.

The control of the driver actuation has to achieve some requirements. Mainly, it has to control brake and accelerator pedal depending on the speed profile of the driving cycle and the vehicle reaction. In case of manual gearshift, the clutch has to be actuated, as well. The government regulations allow a specified tolerance, when following the speed profile of a driving cycle. For example, in the NEDC the driver can depart maximal $2km/h$ from the target speed in both tendencies, below and above (Europäisches Parlament [33]). Additionally, the setting of the parameters in the driver control with regard to different vehicle derivations and driving cycles should be in a simple way.

The primary requirements for the driver actuation model and resulting implications are summarized in table 2.2.

2.2 PHYSICAL SYSTEM: COMBUSTION ENGINE AND VEHICLE POWERTRAIN

In this work, the physical system of combustion engine and vehicle powertrain is described by simulation models. This is different to established methodologies to determine fuel consumption in driving cycles. Usually, the mechanical part of the vehicle powertrain is modeled in

Requirement	Implication
compatible	independent model with simple interconnection
configurable	adjustable by parameters
universal	able to follow various speed profiles

Table 2.2.: Summarizing of key aspects required for the driver actuation model.

a 1D multi-body system. Then, the combustion engine is represented by measured maps or mean value models. The disadvantages of this modeling approach are explained in chapter 1.3 in detail. In order to compensate the mentioned drawbacks, the combustion engine is modeled as well as the vehicle powertrain.

2.2.1 Combustion engine model

Depending on the application and on the required level of detail, different simulation approaches can be used for the modeling of internal combustion engines, as depicted in figure 2.6. With increasing detailing in the modeling of physical properties and processes in combus-

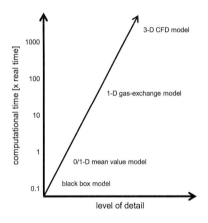

Figure 2.6.: Computational time versus level of detail in modeling (based on Millo et al. [112])

tion engines, the complexity of the models and the computational time rise disproportionately high. The most complete and detailed methodology to describe phenomena of flow dynamics and combustion is the 3D CFD simulation. In the development of internal combustion engines, common fields of application are gas dynamics in components, in-cylinder gas flows like mixture formation, injection spray, and combustion processes. Usually, only specific investigations of individual parts are performed, but no perspective on system level can be provided. For example, modeling the entire air system including intake and exhaust components would result in huge volumes that have to be discretized and thus, in extremely high computational times (Millo et al. [112]).

Complex systems can be described by simplifying the flow properties in 1D fluid-dynamic simulations. The models include networks of tubes connected by orifices or particular subsystems. The 1D transient Navier-Stokes equations for compressible fluids governing the conservation of mass, momentum, and energy of the flow in each element are obtained by finite difference solvers. It implicates constant parameters for the whole cross-sectional area of every individual part without resolution of local effects. This methodology is widely used to determine the behavior of the entire combustion engine like calculations of volumetric efficiency, power output, and fuel consumption. These models show not only a great applicability estimating steady-state operating conditions, but also predictions of transient effects can be accomplished. Further benefits of 1D simulations is its ability for Software in the Loop (SiL) applications, getting more and more important in the development of control systems (Millo et al. [112]).

In order to improve computational speed and trying to keep the accuracy of a 1D model at the same time, so called 0/1D mean value approaches are developed. In purpose to reduce the complexity of a detailed 1D modeling, the main phenomena are still physically described. For example, the network of individual modeled tubes in both intake and exhaust systems can be lumped together in single equivalent volumes. Comparable to a filling/emptying model, the equations of conservation for energy and mass are applied. Belonging to the level of simplifying, the computational time can be reduced significantly. While parameters of the overall system can be predicted, specific information of the gas flow like pressure wave dynamics are not provided any more.

The highest computational speed is achieved by 0D black-box models based on a quasi-stationary approach including experimental data. This methodology using steady-state maps from measurements also supplies the lowest level in detail. It neglects any dynamic effects in the internal combustion engine and thereby, inaccuracies in the simulation of transient operation conditions like e.g. the prediction of response in the turbo-charging system are produced. Additionally, engine maps only work for vehicle derivatives in combination with combustion engines that already have been measured on a test bench. If there are changes in the engine calibration, impacts on power and fuel efficiency have to be estimated.

In purpose to achieve the best compromise between level of detail in modeling and computational requirements, a detailed 1D approach of the combustion engine is used for the investigations in this work. It enables the simulation of the engine characteristics, even if there are adaptations of individual parts or application strategies. Here, only measurements of some operating points covering the interesting parts of the engine operation map are necessary. If no fundamental changes are planned in the engine technology, even measurements from the predecessor engine type can be applied. Once the engine model is parametrized and validated, it can be utilized for various analyses, e.g. with respect to performance or fuel efficiency. Many publications using 1D models to analyze the gas exchange of internal combustion engines in stationary operating points can be found in the literature [22, 52, 83, 121, 140]. This approach can be used to determine maps of the combustion engine specifying its properties (e.g. pressures, temperatures, air flow, and fuel consumption) over the entire range of operation comparable to the process on the engine test bench. In the simulation, particular conditions from the measurement like engine speed, load, and actuator parameters are defined to identify the requested characteristics of gas exchange and combustion after reaching steady-state. The engine model used in this simulation methodology is based on such a steady-state model and is adapted to investigate transient operation.

Summarizing, the proposed engine model represents a good compromise between computational time and quality of results. Compared with 3D CFD, the level of detail is limited in a 1D gas exchange simulation and e.g. no 3D flow effects can be considered. However, it is a detailed model from the viewpoint of low specified engine models in other virtual calibration tools. The modeling level is sufficient to analyze the gas exchange of a combustion engine. Especially, the influence on charging like pressure pulsations in longitudinal direction of the tubes, varying temperatures, and the dynamical characteristics of the turbo-charging can be considered. Last one is important for transient simulations due to turbo lag and response behavior. The use of a 1D-simulation enables the crank angle based calculation of each individual combustion cycle over the entire driving profile in an acceptable period of time. This allows a profound investigation of transient effects and influences of cycle-to-cycle changing parameters by calibration like e.g. ignition angle. From another point of view, it does not offer real-time capability, which would lower the physical model background, decrease the results quality and is not compulsory in a pure simulation environment, respectively. The primary requirements for the engine model and resulting implications are summarized in table 2.3.

Requirement	Implication
detailed	crank angle based calculation
trapped air mass	prediction of gas exchange
power output	additional physical-based (phenomenological) combustion model
transient	moderate computing time

Table 2.3.: Summarizing of key aspects required for the engine model.

2.2.2 Vehicle powertrain model

Due to the aim of analyzing fuel consumption efficiency and the emission output in testing cycles or real-life duty cycles, a correlation between the requested vehicle speed profile and engine characteristics of load and speed is required. This means for the implementation in this simulation methodology, the powertrain of a virtual vehicle with the corresponding kinematics and losses has to be added to consider its dynamics. Within the mechanical model, the generated instantaneous torque of the combustion engine (resolved down to the order of degree crank angle) is converted via rotational powertrain elements into a translational, longitudinal motion of the vehicle. Hereby, it is important to find the right accuracy in the vibration behavior and stiffness of the single components. As driving cycles for fuel consumption analyses are mostly defined via time-dependent target velocities, there are no requirements for lateral dynamic aspects in the vehicle model.

For the purpose of this work, a basically rigid modeling of the powertrain implemented in a 1D-multibody simulation environment is approved. According to high oscillations in the torque signal from the thermodynamic engine model, a damping element has to be added at the interface of the vehicle. Not only kinematics, also the kinetics of the powertrain has to be modeled in the required quality. Thus, the transmission ratios of gearbox and differential have to be considered, as well as the corresponding torque losses. The wheel torque has to be transferred to the road by tire elements including the rolling resistance. Additional power losses dependent on vehicle parameters like the acceleration and aerodynamic resistances have also to be implied in the modeling.

The primary requirements for the vehicle powertrain model and resulting implications are summarized in table 2.4.

Requirement	Implication
power losses	modeling of the powertrain characteristics
vehicle speed	powertrain kinematics
transient	moderate computing time

Table 2.4.: Summarizing of key aspects required for the vehicle powertrain model.

2.3 SIMULATION ENVIRONMENT

After modeling the single part models, the interfaces are connected synchronously in one global model. This synchronization in time is important, if more than one simulation tool is used. In the present case, the importance also arises from the fact that the ECU usually works in different grids in the range of milliseconds and the combustion engine simulation runs in cranktrain synchronized time steps. Additionally, the communication between the models themselves has to be ensured (i.e. the correct transfer of calculated parameters).

The interconnection of the sub-models integrated in the same simulation environment can be arranged easily, since only direct links can be realized (e.g. engine speed and engine torque at the exchange of engine and vehicle). More difficult is the transfer of the parameters from the ECU model to the engine/vehicle model and the other way round (see figure 2.7). Within the proposed simulation methodology, the ECU model described above is coupled to a detailed combustion engine model directly via many actuator and sensor paths. The total number of connections is still lower than in a real vehicle due to implemented model simplifications. However, this method can interchange feed-forward signals and feed-back reactions of the relevant subsystems. On the feed-forward side, the computed actuators from the ECU are transferred to the engine. On the feed-back side, the ECU functions need sensed quantities of actual states from engine and vehicle like in a real car. Additionally, a drivers control is implemented to manage the accelerator and brake pedal dynamically according to the prescribed target velocity of the driving cycle.

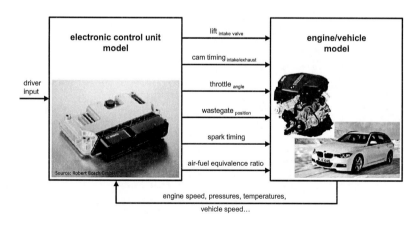

Figure 2.7.: Simulation environment of the coupled overall model

The primary requirements for the coupled overall model and resulting implications are summarized in table 2.3.

Requirement	Implication
synchronous	synchronization in time in all models
communication	correct transfer of parameters

Table 2.5.: Summarizing of key aspects required for coupled overall model.

2.4 FORMATION OF IN-CYLINDER AND TAILPIPE EMISSIONS

In this section, the main sources of in-cylinder emissions in SI engines are introduced and the modeling concept is presented (sections 2.4.1-2.4.3). Basics of the simulation methodology to determine tailpipe emissions are explained in section 2.4.4.

2.4.1 In-cylinder NO_x emissions

In SI engines, NO_x in-cylinder emissions are mainly the product of the oxidation of nitrogen occurring in the intake air at high temperatures. NO_x is a general term for all oxidized nitrogen components NO, NO_2, NO_3, N_2O, N_2O_3, N_2O_4, and N_2O_5. The primary reaction product is nitrogen monoxide NO accounting most for NO_x engine-out emissions (Liesen [96]). Less significant are nitrogen oxide NO_2 and nitrous oxide N_2O. The ratio NO_2/NO is assumed about 2% for SI engines according to Heywood [59].

The sources of nitric oxide emissions during combustion processes can be divided into four main mechanisms (El-Mahallawy and Habik [30]):

1. **Thermal NO_x**: The formation of thermal NO_x at near-stoichiometric fuel-air mixtures can be defined by chemical reactions known as extended Zeldovich mechanism (Heywood [59], Riegler [137]):

$$O + N_2 \; \rightleftharpoons \; NO + N \qquad (2.1)$$

$$N + O_2 \; \rightleftharpoons \; NO + O \qquad (2.2)$$

$$N + OH \; \rightleftharpoons \; NO + H \qquad (2.3)$$

The principal reactions (2.1-2.2), first suggested by Zeldovich, were extended from Lavoie et al. [94] by the third equation considering the effect of oxygen and hydrogen radicals on the formation of NO_x. The reactions mainly take place behind the spreading flame front (post-flame region) in the burned zone. The breaking of the strong N_2 triple bond in equation 2.1 requires huge activation energy limiting the rate of formation with a high temperature sensitivity. Hence, this mechanism can be neglected at temperatures lower than 1600K due to the short retention time of the gas mixture in the cylinder (Riegler [137]). However, the reactions of the oxidation are very fast with low activation energies. In internal combustion engines, the rates of NO_x emissions can be influenced by recycling residual gas to decrease the temperature in the cylinder.

2. **Prompt NO_x**: During combustion, fuel fragmentations can react with N_2 to so called prompt NO_x, also known as Fenimore mechanism (Fenimore [34]). This source is influenced by the radicals of hydrocarbons, mainly contributed by CH and CH_2.

$$CH + N_2 \;\rightleftharpoons\; HCN + N \tag{2.4}$$

$$N + O_2 \;\rightleftharpoons\; NO + O \tag{2.5}$$

$$HCN + OH \;\rightleftharpoons\; CN + H_2O \tag{2.6}$$

$$CN + O_2 \;\rightleftharpoons\; NO + CO \tag{2.7}$$

$$CH_2 + N_2 \;\rightleftharpoons\; HCN + NH \tag{2.8}$$

The mechanism is most prevalent within the flame front in fuel-rich regions and becomes significant at cold-start conditions, but overall, its contribution to the total amount of NO_x emissions is of the order of 5-10% at stoichiometric mixtures (Merker et al. [109], Riegler [137]).

3. **Fuel-NO_x**: Liquid and solid fossil fuels contain nitrogen which reacts with oxygen to NO_x. The contribution of this mechanism to NO_x emissions depends on the concentration of nitrogen bounded in the fuel. The sources of fuel-NO_x are the release of nitrogen-bound compounds into the gas phase during the devolatilization process. The mechanism passes through several reaction steps involving radicals like HCN, NH_3, N, and NH. Today, gasoline for SI engines does not contain fuel-bounded nitrogen compounds and fuel-NO_x does not appear in internal combustion (Merker et al. [109]). Thus, no consideration of this mechanism is required.

4. **Nitrous oxide N_2O**: At low temperatures and lean operation conditions, the mentioned NO_x mechanisms (1.&2.) do not contribute essentially to the engine-out NO_x emissions. Here, the formation of nitrous oxide becomes important and the principle reaction can be defined as (Merker et al. [109]):

$$N_2 + O + N_2 \;\rightleftharpoons\; N_2O + N_2 \tag{2.9}$$

$$NO_2 + O \;\rightleftharpoons\; NO + O_2 \tag{2.10}$$

$$N_2O + H \;\rightleftharpoons\; N_2 + OH \tag{2.11}$$

The primary requirements for the NO_x sub-model and resulting implications are summarized in table 2.6.

Requirement	Implication
modeling	chemical reaction of the NO_x formation
compatible	independent sub-model with simple interconnection
transient	moderate computing time

Table 2.6.: Summarizing of key aspects required for the NO_x sub-model.

2.4.2 *In-cylinder* CO *emissions*

Carbon monoxide CO is a colorless, odorless and tasteless, but highly toxic gas. Generally, the carbon containing in fuel reacts with air to carbon dioxide during complete combustion. CO is an intermediate in the reaction to carbon dioxide and thus, its source is incomplete combustion due to lack of oxygen, low temperature, and short reaction rates (Heywood [59], Merker et al. [109]). During the expansion phase, the formed CO reacts with water vapor to CO_2. This mechanism can be described by the water gas shift equation (Mladenov [115]):

$$CO + H_2O \rightleftharpoons CO_2 + H_2 \qquad (2.12)$$

At stoichiometric and lean operation (lambda ≥ 1) the formation of engine-out CO emissions is very low and nearly does not correlate to the air-fuel ratio. During the combustion process, CO is produced by the dissociation of CO_2 to CO and O_2 and oxidized in the expansion phase afterwards (Eichlseder et al. [29]):

$$CO + \frac{1}{2}O_2 \rightleftharpoons CO_2 \qquad (2.13)$$

The engine-out CO emissions are also widely independent of other engine operation parameters like compression ratio, timing of injection and ignition, and load (except periods of enrichment). A reduction can be accomplished by complete combustion with excess of air or catalytic treatment of the exhaust gases (Sams [143]). The primary requirements for the CO sub-model and resulting implications are summarized in table 2.7.

Requirement	Implication
modeling	chemical reaction of the CO formation
compatible	independent sub-model with simple interconnection
transient	moderate computing time

Table 2.7.: Summarizing of key aspects required for the CO sub-model.

2.4.3 *In-cylinder* HC *emissions*

The formation of HC emissions as a result of fractions of fuel escape burning during the normal flame propagation depends on various sources in spark ignited engines. The literature provides an extensive overview of these effects (Cheng et al. [12]). Figure 2.8 marks the most relevant formation mechanisms in a modern turbo-charged DISI (direct injection, spark ignition) engine described in the following (Dorsch et al. [26]).

Crevices: One of the significant processes that result in unburned hydrocarbons is caused by crevices. Crevices are narrow regions in the combustion chamber where gas can flow into and out, but the flame cannot enter. Unburned gas mixture that remains in these crevices during the combustion progress escapes burning and gets discharged during the exhaust stroke. There are several possible crevice regions in a combustion chamber like cylinder head gasket, spark plug thread, valve seat, and piston ring pack [12]. Their importance depends individually on volumetric and geometric characteristics and component temperature of the engine. Many experimental studies show that the region bordered by piston, compression ring, and liner has the largest volume and accounts most for engine-out HC emissions. Its contribution to total HC reaches values between 40 and 90% according to [9, 12, 61, 113].

During the compression and combustion processes, the mixture of fuel, air, and residual gas is compressed into the volume between liner and piston down to the first piston ring due to the rising cylinder pressure. Since the entrance to this crevice is very narrow and

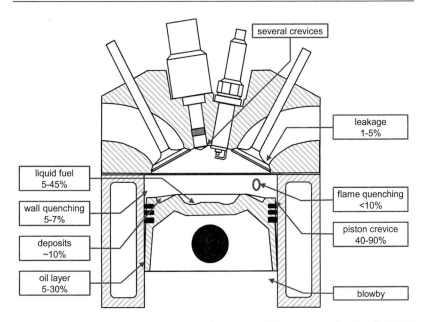

Figure 2.8.: Possible formation mechanisms of engine-out HC emissions referred to [9, 12, 61, 70, 113]

the trapped gas is rapidly cooled near the wall, the flame cannot enter it. In the expansion stroke, the piston crevice mass diffuses partially into the burned mixture while the piston goes down. When it moves up again, the gas layer at the wall gets scraped and it shapes a so called "roll-up vortex" (Merker et al. [109], Trinker et al. [164]). The flows between the top and second rings including blowby effects are more complex, because of essential pressure and temperature drops across the different ring regions [12].

Even if the geometrical volume of the piston crevice is only 1-2% of the complete compression volume, it can trap 4-8% of the mixture mass due to the cold temperature in the crevice [66]. In [9, 158, 160], the influence of geometrical details of the piston affecting the volume of the piston crevice and the texture of the edge at the piston crown (chamfer, discontinuities) is investigated. Another impact on the crevice volume is the expanding of the piston during engine warm-up decreasing its spacing to the liner.

Oil layers: The second important mechanism for engine-out HC emissions is the absorption/desorption of fuel components by the lubricating oil. While the oil layer on the cylinder liner is uncovered by the piston, it gets in contact with the mixture of air, fuel, and residual gas. Hereby, a vapor transfer between the bulk gas and the oil/gas interface in both directions can take place. During compression and combustion, it can be expected that hydrocarbons are absorbed into the oil layer and then diffuse into it. Over the period of expansion and exhaust stroke, the hydrocarbons are desorbed out of the oil film into the cylinder gases and contribute to the engine-out emissions. Concentration profiles of fuel vapor in the gas mixture and in the oil layer during compression (a) and expansion (b) are shown in figure 2.9 schematically.

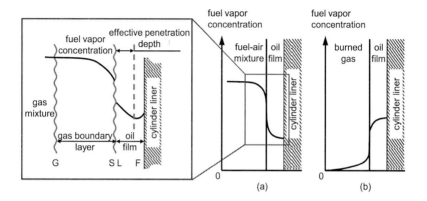

Figure 2.9.: Schematic illustration of the fuel vapor concentration in the different phases according to [16, 37]

Fundamental experiments to quantify the oil layer as a source of 5-30% of unburned HC emissions are conducted in [12, 41, 61]. However, this mechanism depends significantly on the used lubricating oil and engine temperature. The variances in the solubility characteristics of hydrocarbons in the oil result not only from changing temperature and pressure in the cylinder, but also from different thickness of the layer and composition of the oil and fuel.

Deposits: Over the time, deposits build up in the combustion chamber on the valves, piston crown, and walls of the cylinder head. The surfaces of these deposits have many pores where fuel can be absorbed/desorbed. This mechanism is investigated in [155, 160]. The results show a strong dependency on the runtime and operation condition of the engine and the used fuel. Overall, only a small effect on the engine-out HC emissions of around 10% is assumed [12].

Wall quenching: During burning, the flame front extinguishes a small distance from the cylinder walls due to their cooled temperature leaving a thin layer of unburned gas mixture close to the liner (called quench layer) [55]. At warmed-up engine condition, most of this unburned mixture diffuses into the hot burned gases in the cylinder and burns up finally during expansion. Thicker quench layers occur under cold start operation and a complete burn up cannot be ensured. Then, this mechanism can be an additional source of 5-7% of engine-out HC emissions [12, 70].

Flame quenching: Partial burning effects mean a flame extinguishment prematurely and can cause very high unburned hydrocarbons engine-out emissions. These can occur under critical burning conditions like the engine operation close to its limits of dilute (high rate of residual gas) and lean mixture. This mechanism is only important for transient engine operating conditions with high exhaust gas recirculation. Its contribution is assumed far below 10% [61].

Liquid Fuel: Especially in DISI engines, the injected fuel can impact the cylinder walls and the piston crown. Usually, this is compensated adjusting the injection by pressure and time.

It has to be considered primarily during engine warm-up. If the surfaces in the combustion chamber are still cold, only a fraction of the injected fuel will evaporate. Remaining liquid fuel drops on the walls will be absorbed by the oil layer, deposits, or stored in crevices. However, to achieve a stoichiometric air-fuel ratio in the exhaust gas and to compensate the power loss, more fuel than normally required with complete vaporization has to be injected. In the ECU, this is corrected by scaling the requested fuel mass using a multiplier depending on the engine temperature (see 3.1). It is difficult to quantify this source, because of its combination with other storage mechanisms and due to its strong dependency on the used technologies of injection and charge motion. Consequently, this effect is expected in a wide range between 5 and 45% of engine-out HC emissions [70].

Leakage: In older engines, another source of engine-out HC emissions can be exhaust valve leakage. Hereby, unburned gas mixture flows into the exhaust manifold even the valves are fully closed. Usually, the amount of hydrocarbons is only marginal (rates about 1-5%) and can be negligible [12]. Additionally, today's production techniques allow narrow tolerances resulting in a fitting of the valves without leakage.

In-cylinder oxidation: Oxidation of unburned hydrocarbons has a strong influence on the final amount of HC emissions emitted by the engine. The unburned gas mixture can oxidize within the cylinder or the exhaust manifold before it arrives the three-way catalyst. If the temperature of the burned gas is hot enough (above 1000K) and sufficient oxygen is provided, the unburned fuel will burn up and the amount of HC emissions will be reduced significantly [12]. Studies of in-cylinder oxidation [126, 150, 164] indicate for this process an oxidation rate of 40 up to 70%. Beneath the oxidation temperature, the time that is available for the burn up has a significant influence on the amount oxidized. Cheng et al. [12] suggest that oxidation also depends on the source of HC emissions. Mechanisms producing unburned fuel-air mixture (crevices, quenching) are mentioned to have higher oxidation rates than sources emitting fuel vapor (oil layer, deposits).

Depending on the used engine technology and measurement techniques, the influence of all mentioned mechanisms to the fraction of unburned fuel emitting the cylinder after combustion can differ in the investigations found in literature. Based on the literature review above, the different effects are quantified in figure 2.8 to identify the key mechanisms of HC engine-out emissions. Thus, the main sources are effects of piston crevice volume and oil layer absorption/desorption. Additionally, a strong temperature dependency is indicated in the formation of in-cylinder HC emissions. Hence, the primary requirements for the HC model and resulting implications are summarized in table 2.8.

Requirement	Implication
modeling	formation mechanisms of crevice and oil layer
effects	considering temperature dependency
compatible	independent sub-model with simple interconnection
transient	moderate computing time

Table 2.8.: Summarizing of key aspects required for the in-cylinder HC emissions model.

2.4.4 Tailpipe emissions

The limits for pollutants regulated by law are always based on the amount of emissions emitting at the end of the tailpipe. In order to achieve a complete virtual environment to predict vehicle emissions, a simulation model considering the exhaust aftertreatment of a SI-engine has to be added to the calculation of engine-out emissions. As described in chapter 1.3, the pollutants of the vehicle investigated in this work are converted by a three-way catalyst. The basic construction of a TWC can be separated into three parts:

Core: The catalyst support or substrate is a temperature-stable honeycomb structure. Usually, it is composed of ceramic, but it can consist of metallic foils as well. The monolith is designed with a multiplicity of parallel thin-walled tubes to provide a large surface area. The number of channels is defined by cells per square inch (cpsi). A high reactivity (surface area) can be attained by increasing the cell density or reducing the wall thickness, but this also effects the flow resistance and mechanical stability of the catalyst. Current TWC's range from 200 to 1200cpsi with typical wall thickness between 50 and 254μm (Hagelüken [53]).

Washcoat: The washcoat is applied to the core carrying the catalytic materials such as aluminium oxide (Al_2O_3) and silicon oxide (SiO_2), a mix of precious metals, and components for the storage of oxygen like cerium(IV) oxide. It is very porous and forms an irregular surface with a high roughness to maximize the catalytic active surface area (Nijhuis et al. [123]). Nowadays, catalysts can be coated with multiple washcoat layers including different compositions. Its thickness ranges between 10 and 200μm due to its nonuniform randomly distribution in the cell (Holder [63]). The catalyst mechanisms are affected by the application of precious metals like platinum (Pt), rhodium (Rh), and palladium (Pd). Platinum and palladium are preferred for oxidation, whereas rhodium is qualified for reduction of nitrogen oxides (Mladenov [115]). Due to high costs, platinum is replaced by palladium and a composition with rhodium is commonly applied in modern catalyst converters.

Canning: The ceramic core is surrounded by a metallic canning. The metal body protects the catalyst by outside influences like dirt, water, and mechanical damages. Between the monolith and the canning, a thermal insulation prevents a fast cooling of the catalyst.

The task of a TWC is the chemical conversion of toxic pollutants to innocuous emissions like carbon dioxide (CO_2), water (H_2O), and nitrogen (N_2) with the help of simultaneous oxidation and reduction reactions (Holder [63]):

1. Reduction of nitrogen monoxide and carbon monoxide to carbon dioxide and nitrogen:

$$2NO + CO \rightarrow CO_2 + \frac{1}{2}N_2 \qquad (2.14)$$

2. Oxidation of carbon monoxide to carbon dioxide:

$$2CO + \frac{1}{2}O_2 \rightarrow CO_2 \qquad (2.15)$$

3. Oxidation of unburned hydrocarbons to carbon dioxide and water:

$$C_mH_n + (m + \frac{n}{4})O_2 \rightarrow mCO_2 + \frac{n}{2}H_2O \qquad (2.16)$$

The aim of the catalyst model is to simulate the exhaust aftertreatment of the main pollutants considering the chemical mechanisms. For this purpose, the process of the emission conversion in a TWC can be classified into seven steps (Mladenov [115]):

1. diffusion of the educts from the exhaust gas flow to the washcoat surface in the channels.

2. further diffusion of the educts into the pores of the washcoat.

3. adsorption on the surface area.

4. chemical reaction on the catalytic active surface area.

5. desorption of the products from the surface area.

6. diffusion of the products out of the pores.

7. diffusion of the products from the washcoat to the exhaust gas flow.

The primary requirements for the tailpipe emissions model and resulting implications are summarized in table 2.9.

Requirement	Implication
modeling	chemical mechanisms for the conversion of the main pollutants
reproduction	considering the conversion process in a TWC
compatible	independent model with parametrized interconnection

Table 2.9.: Summarizing of key aspects required for the tailpipe emissions model.

3

MODELING

In this chapter, a calibration environment combined with a thermodynamic engine model including a predictive combustion sub-model and an appended virtual vehicle powertrain is developed to simulate various driving cycles. Hereby, the seamless coupling of different sub-models, as well as the preparation of the combustion engine model towards dynamic operation is the main challenge. In the following sections, the modeling of the coupled approach with respect to the requirements from chapter 2 is described in detail. The modeling is presented starting with the control system of ECU (3.1) and driver actuation (3.2), followed by the physical system engine (3.3) and vehicle (3.4), and summarized by the coupled environment in 3.5. Additionally, the determination of in-cylinder (3.6) and tailpipe emissions (3.7) is explained and validated by steady-state measurements.

3.1 ECU MODEL

One way to simulate a virtual engine control accurately is the straight adoption of the source code from the real ECU. However, these Software-In-The-Loop tests generate high implementation costs due to the essential supply of all input parameters by virtual sensor signals. Additionally, an overview of the control system and the correlation of individual modules and functions can be provided only complicated. The requirements from chapter 2.1.1 demand a transparent and modular modeling. Taking the advantage of the model based approach (presented in 2.1.1) in the development of ECU's, the main modules and corresponding functions can be adopted independently. Furthermore, simplifications and modifications due to simulation needs can be adapted. Here, the challenge is to find a well balanced way between reducing the complexity and still providing all important connections and interfaces with regard to engine calibration demands.

The chosen simulation tool is *Simulink* from *The MathWorks Inc.* (MathWorks [103]). It has an interactive, graphical development environment to map dynamical functions and control systems. It provides an adequate library containing mathematical and logical functions, elements for signal routing or constant parameters and look-up tables. The tool is based on *Matlab* (from *The MathWorks Inc.* as well) enabling pre- and post-processing controlled by self-coded programs. Actually, there is going on a change to model based development of functions for the engine control. Signal paths are created and tested directly in *Simulink* and compiled into the source code by an automated process (e.g. Real-Time Workshop Embedded Coder (RTW-EC)), afterwards. No such software implementation models are provided for this work, so far, but this modeling approach enables easier adaptions of further developments in the future.

The ECU model is divided into discrete modules with sub-elements based on the structure of the real functions. This allows a hierarchical modeling with a tree-like structured user interface. Signal paths are visible and dependencies or interactions of particular functions can be retraced easily. A strong simplification can be achieved by concentrating of only relevant calculations. Usability for calibration work is ensured by implementation of the original parametrization into the model. With the help of a BMW internal add-on tool for *Matlab*, the original ECU calibration file is translated to a MAT-file object (internal standard format for *Matlab*). This requires a fixed definition of the naming for the variables in the objects for constant parameters or look-up tables and has to be observed in the modeling.

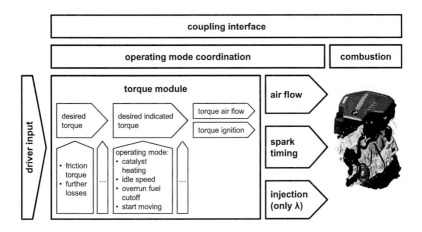

Figure 3.1.: Main modules of the architecture level in the ECU model

Figure 3.1 illustrates the highest level of the ECU model with the main modules focusing on the torque functions. The structure follows very close the real engine management containing the main functions described in chapter 2.1. In the following, the main functionality of each module is introduced and modifications in the modeling are presented.

Coupling interface

In the coupling interface, all relevant parameters for the actuation of engine and vehicle are collected and transmitted. Sensed signals from the physical models are processed and linked to the corresponding functions. The coupling process is conducted by a linking object designed to invoke the engine simulation solver and provided by *Simulink*.

Coordination of operation modes

The different engine operation modes are controlled in the module coordination of operation modes. In this work, the special operation modes of idle speed, catalyst heating and overrun fuel cut-off are necessary in the driving cycles. Idle speed is a specific operation situation for the engine, since no load is required by the driver. Without control, this would result in no torque request. Due to the engine friction, the ECU has to anticipate an engine load. This idle load can vary depending on engine temperature and auxiliary torques like air conditioning, power steering systems, and electrical charging. The ECU controls the engine speed directly by fuel injection, ignition timing, and gas exchange. The control function for the activation of catalyst heating can be simplified, since the time period for a defined engine/catalyst combination correlates during a certain driving cycle. Thus, this information is taken from measurements in this simulation methodology. However, the torque request regarding catalyst heating is calculated separately in the torque module. The control of overrun fuel cut-off has its own coordination of operation mode. Since this module is complex due to different intermediate transition modes, it is reduced to only two states (switch on/off). The simplification should not have an impact on the operation in the selected driving cycles, but the additional modes are important for driving comfort.

Torque module

In the torque control, the desired torque of the combustion engine has to be evaluated with respect to external torque requests from e.g. the managements of driving dynamics and gear box. These requests are disregarded in this work, since only functions of the engine control are modeled. Driving dynamics can be ignored due to the only longitudinal driving cycles used for investigations of fuel consumption or emissions. Here, the control of the gear box can have a not insignificant influence, but mainly for automatic transmission. The torque module can mainly be described by three calculation steps [21, 140, 170]:

Conversion of the gas pedal: With the help of a progression map, the position of the gas pedal is converted into a relative desired torque in accordance to the actual engine speed. There can be more than one progression map for the accelerator pedal in a vehicle, e.g. depending on different driving modes for comfort, dynamic, and fuel saving driving. Here, the driver can change the response behavior of the vehicle on his demand of driving dynamics. The relative torque TQ_{req_rel} gets interpolated between the drag (TQ_{min}) and full load (TQ_{max}) torque of the engine. This results into the absolute requested torque TQ_{req} at the cranktrain defined by

$$TQ_{req} = TQ_{req\_rel} \cdot (TQ_{max} - TQ_{min}) + TQ_{min} \qquad (3.1)$$

Load-reversal damping and anti-vibration function: Fast changes of the acceleration pedal position are controlled by the loadreversal damping (positive) and dashpot function (negative gradient) to delay dynamically increases and decreases of the engine torque. Several ramp functions smooth the desired torque to avoid uncomfortable vibrations in the powertrain. Additional intervention is achieved by the anti-vibration function. It superimposes an inversely phased torque to the vibrations due to changes in the ignition.

Coordination of torque: The functional module of the torque coordination is presented in figure 3.2. The desired torque due to the drivers request has to be adjusted regarding internal and external torque requests. Therefore, the indicated torque of the engine (TQ_i) is calculated based on the required and corresponding drag torque TQ_{min}:

$$TQ_{i\_req} = TQ_{req} - TQ_{min} \qquad (3.2)$$

The drag torque includes all torque losses like friction losses and auxiliary torques of e.g. oil pump, electric generator, and air-conditioning compressor. Friction losses depend on engine speed and temperatures of oil and coolant fluid. External offsets can be defined by controls of the gear box and driving dynamics (e.g. TCS), and the energy management (also included in demands of idle speed).

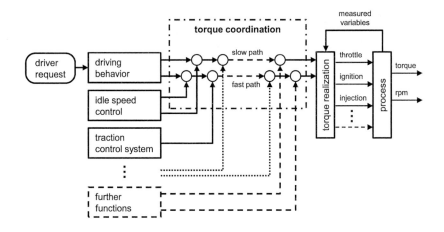

Figure 3.2.: Functional module of the torque coordination [36, 68, 69]

The torque requests have different priorities at the time of progress. Hence, the calculation is divided into two paths. In the slow charge path, the torque $(TQ_{i\_req\_c})$ for the control of the air mass is calculated. In contrast, the fast spark path $(TQ_{i\_req\_s})$ is for the actuation of the spark timing. The separation of the load control is founded on the different response behavior of these two systems. In purpose of changing the load by air, the right charge has to be filled into the cylinder. Compared to the calculating time at 100ms intervals of the ECU, the gas exchange is limited by the dynamics of the intake and exhaust system. However, the spark timing can be modified synchronized to the angle of the cranktrain without nearly any time lag, i.e. within one working cycle. The change of last one is often combined with additional fast (crank angle synchronized) parameters like air-fuel equivalence ratio and injection timing (Mencher et al. [107], Nietschke et al. [122]).

Air flow

In the air module, the desired torque of the air path gets converted into the actuation parameters for valves, throttle and wastegate (see figure 3.3). The air mass is continuously adjusted by a control loop of matching requested to actual cylinder charging. Therefore, sub-functions of the load detection and load control are required (Nijis et al. [124], Roithmeier [140]):

Load detection: The aim of the load detection is to determine the actual amount of fresh air, which is provided for the combustion into the cylinders at every working cycle.

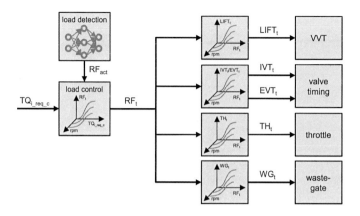

Figure 3.3.: Functional module of the air system according to Roithmeier [140]

Usually, this is done dividing the actual air flow ($\dot{m}_{air,cyl\_act}$) by the air flow under standard conditions ($\dot{m}_{air,cyl\_norm}$). The result is the relative cylinder charging (RF_{act}):

$$RF_{act} = \frac{\dot{m}_{air,cyl\_act}}{\dot{m}_{air,cyl\_norm}} \qquad (3.3)$$

The actual mass flow rate of intake air gets sensed by a Mass Air Flow sensor (MAF) (a common type is the hot wire sensor). This is generally installed in front of the throttle due to a low level of turbulence in the flow at this location (shown in figure 3.4). However, neg-

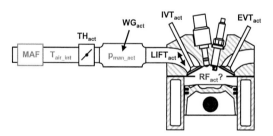

Figure 3.4.: Load detection in the air system with its sensors and actuation parameters [105]

ative impacts according to this position can appear, especially in situations of dynamic load changes. Discontinuities in the calculation of the cylinder charging can occur by phenomena of vibrancies in the intake manifold, residual gas content in the cylinder, and scavenging.

Consequently, in modern ECU's, engine load is calculated by modeled functions. Depending on the used engine technologies, RF_{act} can be defined by engine speed, valve timing, intake valve lift, and positions of throttle and wastegate. Additionally, measured inputs of present load sensing models are values from the air flow sensor, pressure in the intake manifold (p_{man_act}) and position of the throttle (TH_{act}). However, calculations based on cylinder pressure are also feasible [47, 71]. A pure physical based modeling is complex, although the computing has to be in real-time. An alternative solution is a map based function or a combination of both. If there is used a modeling with more than four input parameters (e.g. in the case of full variable valve actuating), the calculation of the engine load by maps is complicated and can only be achieved with high calibration and computing costs. Here, a common way is to use artificial neural networks (ANN) [89, 105, 124]. These are computational models interconnecting virtual neurons that can compute a physical process with multidimensional input parameters and unknown functional correlation by an optimization technique consisting of learning and pattern recognition. In ECU's, static ANN's can be used, as well as online adaptive ones that learn in operation by actual measurements (Lichtenthäler [95], Nelles [120]).

With the help of ANN's, the engine control can be adapted to each individual engine due to tolerances in the production of the engine, vehicle or sensors. However, the aim of this work is to represent a common standard engine. Without ANN's, the complexity of the ECU model is reduced to maximize transparency. The difficulty is the replacement of the ANN by adequate functions. However, using a detailed engine model provides the advantage of simulating the gas exchange in the cylinder. Thus, the amount of fresh air in the cylinder can be detected and transmitted to the ECU model. According to equation 3.3, the relative cylinder charging (RF_{act}) can be calculated with the simulated air flow ($\dot{m}_{air,cyl\_act}$), as well. The air flow under standard conditions ($\dot{m}_{air,cyl\_norm}$) is determined by a conversion parameter dependent on engine speed and defined by the engine calibration.

Load control: In the load control, a continuous adjustment of requested to actual air mass is required (Roithmeier [140]). Therefore, the required torque for the air path ($TQ_{i\_req\_c}$) is converted into a target cylinder charging (RF_t) by considering the actual RF_{act}.

Engine actuation: The target cylinder charging RF_t is essential to control the actuator parameters of intake valve lift, valve timing, and positions of throttle and wastegate. These are mainly defined by maps depending on engine speed and target cylinder charging. The resulting target positions/timings of each component are adjusted by several sub-functions and then transmitted to the engine. Small modifications are adapted to the calculation of the wastegate position for the proposed engine model in this work.

At naturally aspired engine loads, the wastegate is closed to achieve higher speeds of the turbo rotor and to enable faster response characteristics in case of sudden load requests. Above this partial-load range, the boost pressure is controlled by opening the wastegate considering surge and choke line. In this actuation module, a relationship between requested load pressure at the intake and the pressure drop across the turbine at the exhaust is developed. In dependency of the desired pressure in the intake manifold and ambient pressure, a pressure ratio at the compressor is determined. This is converted to a pressure ratio at the turbine by a stationary balance of power. Compared with the current state, this results in an actuation parameter for the electrical adjustment of the wastegate. In the context of the present work, the last step is omitted and only the required boost pressure of this calculation is taken. The wastegate is not modeled as valve with an electric actuated flap in the proposed engine model, but as orifice connection with variable opening diameter. Thus, the actuation is controlled directly in the engine model in accordance to the requested boost pressure of the virtual ECU.

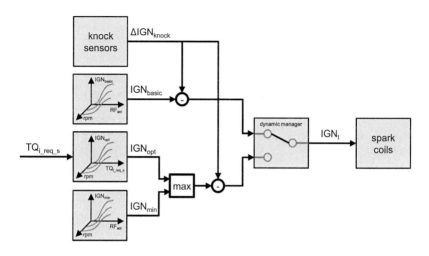

Figure 3.5.: Functional module of the ignition system according to (Roithmeier [140])

Spark timing

Additionally to the air and fuel amount in the cylinder, the spark timing has an important role to control the power output of a combustion engine. The combustion efficiency can range between effective fuel consumption and maximizing of torque output. However, the modification of ignition is limited by knock (spark occurs too early in the compression stroke), exhaust temperature, and complete combustion of the air/fuel mixture (unsteady running). These functional claims are controlled by the spark module in the ECU (illustrated in figure 3.5) (Roithmeier [140]):

Basic ignition angle: Under normal conditions (i.e. steady-state driving), a basic spark timing (IGN_{basic}) is applied. It is calibrated to a center of combustion mass with regard to low fuel consumption. But if a high load is requested, it changes automatically from effective fuel consumption to early timings limited due to engine protection. Knocking is already considered in these maps.

Optimized ignition angle: In some cases, the basic spark timing has to be adapted on the driver request. Fast responses in engine torque can be achieved by changing the ignition to earlier or later timings. These interventions in ignition are required, e.g. for the control of idling speed, several functions of driving dynamics, and special operating modes of the engine like catalyst heating. The optimized ignition angle (IGN_{opt}) depends on the torque request by the driver (TQ_{req}).

Dynamic manager: The dynamic manager controls the requests of fast torque interventions concerning ignition angle. It switches between the fuel-efficient optimized ignition angle and the required one due to torque demands. This manager is strongly simplified in the presented ECU model, since in the investigated driving cycles such fast changes requested by e.g. the driving dynamic control are not required. The result is the target ignition angle (IGN_t) actuating the spark coils in the engine.

Minimal ignition angle: As mentioned above, it is important not to exceed the combustion boundary to burn the injected fuel completely. The resulting high temperatures in the exhaust system can damage its components. In some extent, this effect is used during catalyst heating. The limitation in late spark timing is described by the minimal ignition angle (IGN_{min}). Corresponding to the declaration that all ignition angles before Firing TDC are positive and these after TDC are negative, the latest possible spark timing is also the minimal ignition angle.

Knock control: Even, knocking is already considered in the maps of the basic ignition angle, a function for knock control is added in the spark module. This control is essential, since knocking is a dynamically phenomena that cannot be predicted precisely. The knock control detects critical pressure oscillations in the cylinder to protect the engine. This can effect essential changes for the ignition. If pressure fluctuations are detected in the cylinders, this function can change the ignition angle to later ones by subtraction of a calculated extension (ΔIGN_{knock}) and phases out smoothly. However, this module is not modeled in this work due to its complexity and to its operation mainly at full load of the engine. The latter is not reached at the driving cycles in combination with the engine used in this research.

Injection control

In order to guarantee a clean and effective combustion, the quantity of fuel to inject has to be determined accurately by the ECU. Nowadays, all vehicles with SI-engines are equipped with a 3-way catalytic converter to reduce emissions. This requires a stoichiometric mixture controlled with the help of a so called lambda oxygen sensor measuring the amount of residual oxygen in the exhaust gas. Combined with the load detection, the mass of fuel to inject into the cylinder can be calculated in the ECU.

The module of injection can be ignored, because it consists mainly of an adaption (ANN) with regard to injector drifts and fuel quality. These are not considered in this work due to the aim representing a common standard engine (same demand as for the ANN's of the load detection). Assuming a homogeneous engine operation with stoichiometric air-fuel ratio allows the evaluation of the required fuel mass depending on the air charge in the cylinder. At cold start conditions of the engine, a relative offset as function of the cooling temperature is multiplied to the calculated mass of fuel (for stoichiometric mixture) to compensate the wetting of cold cylinder liners. This compensation is required to achieve a stoichiometric mixture in the exhaust gas for an optimal conversion rate of the TWC. Thus, the additional mass of fuel cannot be seen in the lambda signal which is important for considering fuel consumption in driving cycles. In this simulation method, this parameter is used in the post-processing to determine fuel consumption and raw emissions at warm-up operations.

3.2 DRIVER ACTUATION

A driver actuation is needed to follow the vehicle speed profile of the considered driving cycle. This is achieved by the control of brake, accelerator, and clutch (for manual gearshift) pedal. According to the requirements of chapter 2.1.2, the driver actuation is integrated into the ECU model as independent module and its calibration regarding different vehicle types and driving cycles is managed by variable parameters. The driver control includes a combination of a feed-forward and a feedback control, as presented in figure 3.6:

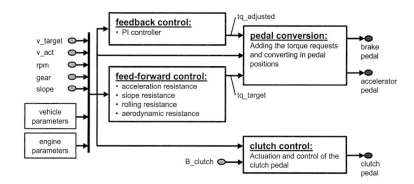

Figure 3.6.: Main modules of the driver actuation

Feed-forward control: In the open-loop function, the required engine torque to achieve the requested vehicle speed profile is calculated reversely. The driving force is determined by separating the main vehicle losses into acceleration, slope, rolling, and aerodynamic resistance. According to the target vehicle speed and actual gear, the required total power is identified and converted to a torque request with respect of axle and gear ratios. The evaluation of the driving resistances is specified in section 3.4.

Feedback control: With the help of a proportional-integral (PI) controller, the relative torque deviation based on the actual vehicle speed discrepancy is determined in the closed-loop part. The absolute torque demand is calculated due to the actual engine speed and torque characteristics of full load and overrun mode. Additionally, a trigger as function of vehicle stops resets the integrator of the PI controller to avoid integral windup.

The torque request of the feed-forward control (tq_target) is weighted as function of the target vehicle speed and added to the result ($tq\_adjusted$) of the closed-loop module. The total required torque is converted to a relative one and translated to a pedal position of accelerator and brake respectively. If the vehicle stops and the target speed is zero, an assumed constant brake pedal of 20% is actuated.

The actuation of the clutch pedal is required for vehicles with manual gearshift. According to the selected driving cycle, the time periods of disengaging the clutch are predefined and provided as input parameter (B_clutch). However, the direct switch between 0 and 100% pedal position in this profile has to be adapted for the usage in the ECU and vehicle models. Especially during drive away situations, the disengaging has to be smooth avoiding engine stall. When the driver starts to remove his foot from the clutch pedal, it is detected in the

43

ECU and an additional torque is requested to support driving off. For this purpose, the clutch signal is filtered only during disengaging and the clutch pedal was actuated for at least $2s$. Thus, fast and direct actuation is still feasible during shifting and engaging the clutch.

3.3 ENGINE MODEL

3.3.1 *Modeling*

The thermodynamic processes of the combustion engine are implemented by a 1D gas exchange model using the commercial software *GT-SUITE* [40]. The calculation of the inherent gas dynamics is based on 1D transient Navier-Stokes equations for compressible fluids. For example, the pipe components are implemented as tubes with variable cross-sectional area A_{cs} simplifying the computation of the flow into the characteristic streamwise direction x to following equations (Merker et al. [109]):

- mass conservation:

$$\frac{\partial \rho}{\partial t} + u\frac{\partial \rho}{\partial x} + \rho\frac{\partial u}{\partial x} + \rho u\frac{1}{A_{cs}}\frac{\partial A_{cs}}{\partial x} = 0 \tag{3.4}$$

- momentum conservation:

$$\frac{\partial (\rho u)}{\partial t} = -\frac{\partial (\rho u^2 + p)}{\partial x} - \rho u^2\frac{1}{A_{cs}}\frac{\partial A_{cs}}{\partial x} - \frac{F_{friction}}{V} \tag{3.5}$$

- energy conservation:

$$\rho\frac{\partial E}{\partial t} = -\frac{\partial (\rho u E + up)}{\partial x} - (\rho u E + up)\frac{1}{A_{cs}}\frac{\partial A_{cs}}{\partial x} + \frac{\dot{Q}_{wall}}{V} \tag{3.6}$$

This neglects the internal friction and thermal conduction in the fluid as well as the force of inertia. Friction is only considered at the tube walls as

$$\frac{F_{friction}}{V} = \frac{\lambda_{friction}}{2d}\rho u\,|u| \tag{3.7}$$

with the dimensionless friction coefficient $\lambda_{friction}$ and diameter d of the tube. The heat conduction at the tube walls $\dot{Q}_{wall}$ is considered by Fouriers law (3.8) with heat transfer coefficient α_{wall}, wall temperature T_{wall}, and surface area A_{wall}, respectively:

$$\dot{Q}_{wall} = \alpha_{wall}A_{wall}\left(T_{wall} - T\right) \tag{3.8}$$

The parameters pressure, temperature, and flow velocity of the one-dimensional flow in the pipe component can be evaluated numerically by solving the conservation equations (3.4 to 3.6) and using the ideal gas equation

$$pV = mRT \tag{3.9}$$

with the specific gas constant R. Therefore, the tube is discretized into small sections.

Air intake system

In the engine simulation, the entire intake/exhaust gas path of a turbo-charged combustion engine is modeled from air intake to tailpipe by physical objects (pipes, boxes, junctions, orifices, throttles, etc.). The intake starts with the air filter at ambient conditions, followed by the compressor of the turbo-charger linked mechanically to the turbine by a shaft. In the 1D environment, the turbo-charger is modeled by maps consisting of measurements of charger speed, pressure ratio, mass flow and efficiency. These measurements are performed on specialized flow-analysis test benches investigating mass flow, pressure ratios, and adjacent temperatures. The generating of the maps is then supported by 3D CFD simulations [141, 142]. This process is already implemented in the development of new engine series and enables an easily adoption to this simulation methodology. Attention is demanded to the maps describing also operation ranges at very low speeds, usually occurring at low loads in driving cycles.

In order to increase the effective power of a turbo-charged engine, the compressed air has to be cooled down after its warming due to the pressure raise. In the intercooler, the temperature in the intake path is cooled by the air flowing against the front of the vehicle. This results in a strong dependence on the vehicle speed. Usually, this is emulated by varying the performance of the air blower on dynamometers, while a constant air flow can be assumed on engine test benches. In the engine simulation, the intercooler is modeled as heat exchanger consisting of many tiny, tubes with defined wall temperature and a high heat transfer coefficient. By means of specifying the temperature, variable oncoming flows against the intercooler and engine components at different vehicle speeds can be considered. In this methodology, the tube walls are conditioned by a measured profile of the intake air temperature in front of the throttle (T_{air_int}) during the investigated driving cycle. Analyses of various environment specifications show a significant impact on the prediction of fuel consumption. Only small deviations of 7-10$mbar$ at the intake pressure result in a difference of 0.1% CO_2 emissions at the end of the NEDC, for example.

After the intercooler, the pressure in the intake manifold is controlled by the throttle. It is actuated by the ECU providing the opening angle in percent. In contrast to the real engine, here the throttle is modeled by an orifice connection with a fixed diameter (taken from the real throttle). The flow constriction according to the flap position (reducing the geometrical cross-sectional area of the tube) and further restrictions resulting from friction losses due to circulations at the flap are implemented by specifying variable discharge coefficients. Therefore, a correlation between effective $\dot{m}_{air,eff}$ and maximum theoretical $\dot{m}_{air,max}$ air flow can be defined as (Merker et al. [109]):

$$\mu_c = \frac{\dot{m}_{air,eff}}{\dot{m}_{air,max}} \tag{3.10}$$

In this modeling approach, the actuation requires a translation of the opening angle into the corresponding coefficient. According to the discharge equation of Saint-Venant and Wantzel (Merker et al. [109], Schade et al. [145]), the velocity of a steady flow at an adiabatic throttle can be expressed by the isentropic relation

$$v_1 = \sqrt{\frac{2\kappa}{\kappa - 1} \frac{p_0}{\rho_0} \left[1 - \left(\frac{p_1}{p_0} \right)^{\frac{\kappa-1}{\kappa}} \right]} \tag{3.11}$$

with the density ratio

$$\frac{\rho_1}{\rho_0} = \left(\frac{p_1}{p_0}\right)^{\frac{1}{\kappa}} \tag{3.12}$$

The static pressures in front of (p_0) and behind (p_1) the throttle can be measured. The calculation of the density ρ_0 is based on the ideal gas equation:

$$\rho_0 = \frac{p_0}{RT_0} \tag{3.13}$$

The mass flow is defined as

$$\dot{m} = A_1 \rho_1 v_1 \tag{3.14}$$

Inserting equations 3.11 and 3.12 into 3.14 the theoretical maximum mass flow results in

$$\dot{m}_{air,max} = A_1 \sqrt{p_0 \rho_0} \sqrt{\frac{2\kappa}{\kappa - 1} \left[\left(\frac{p_1}{p_0}\right)^{\frac{2}{\kappa}} - \left(\frac{p_1}{p_0}\right)^{\frac{\kappa+1}{\kappa}} \right]} \tag{3.15}$$

According to equation 3.10 and considering 3.13 the discharge coefficient can be expressed as function of measured parameters like actual air mass flow, pressures and temperature:

$$\mu_{TH} = \frac{\dot{m}_{air,eff}}{A_{TH,max} \sqrt{p_0 \frac{p_0}{RT_0}} \sqrt{\frac{2\kappa}{\kappa-1} \left[\left(\frac{p_1}{p_0}\right)^{\frac{2}{\kappa}} - \left(\frac{p_1}{p_0}\right)^{\frac{\kappa+1}{\kappa}} \right]}} \tag{3.16}$$

This enables the determination of the discharge coefficient for each operation point of the engine. Its correlation to the throttle position is shown in figure 3.7. The measured data (black

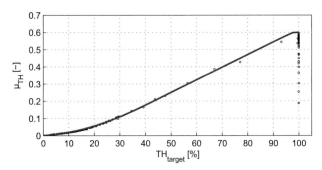

Figure 3.7.: Throttle actuation by discharge coefficient as function of throttle position

dots) belongs to an entire engine map at stationary operation. At turbocharged loads, the boost pressure is controlled by the wastegate opening the throttle completely. Hence, various values are evaluated for the discharge coefficient at a position of 100%. The red line represents the implemented function to convert the throttle position of the ECU into the actuation (discharge coefficient) of the orifice connection. The validation of the manifold pressure in the driving cycle has confirmed this approach and its transfer to transient operation. Results are shown in chapter 4.

Engine cylinder

The essential element of the engine model is the cylinder. The air charge is depending on the predicted gas dynamics in the cylinder influenced by the valve operation. The actuation of the valves is modeled by profiles of the opening stroke versus crank-angle and is controlled by parameters of maximum lift and timing at the intake and only timing at the exhaust. Comparable to the definition at the throttle, the air flow into and out of the cylinder is strongly affected by the definition of the discharge coefficients of the intake and exhaust valves. In this modeling approach, characteristic curves of these coefficients are specified versus valve lift separated for both flow directions, forward and reverse. The information is gained from measurements conducted on purpose-built test benches (Dorsch [22], Merker and Schwarz [108]). According to former analyses in the literature [22, 48, 104, 140], the measured characteristics have to be adapted for full-variable valve actuation due to inconsistency effects between the measurement methodology and real engine operation. Difficulties in calibrating the discharge coefficients are presented in chapter 4.3.2.

According to the amount of fresh air, the mass of fuel to inject is determined. The air mass m_{air} is detected after closing of the intake valves considering the residual gas content. Usually, the value for residuals is provided at the beginning of the actual working cycle (e.g. at gas exchange TDC) corresponding to the previous one. In this work, an adapted determination is added to achieve more accuracy in the validation of fuel consumption. It includes the virtual sensing of the actual oxygen concentration $X_{O_2^{cyl}}$ and trapped mass m_{cyl} in the cylinder as

$$m_{air} = m_{cyl} \frac{X_{O_2^{cyl}}}{X_{O_2^{air}}}; \ X_{O_2^{air}} = 0.233 \tag{3.17}$$

Based on the defined air-fuel ratio of the ECU, the duration of injection is calculated and thus, the fuel mass. Finally, a homogeneous fuel-mixture generation is assumed for the combustion prediction.

In purpose of predicting fuel consumption and engine-out emissions, it is essential to achieve a sufficient quality in the identification of the combustion process. The pressure characteristics in the cylinder is determined by a thermodynamic cycle calculation. In the cylinder system, the mass balance without considering any leakage at the piston rings can be written as

$$\frac{\partial m_{cyl}}{\partial t} = \frac{\partial m_{int}}{\partial t} + \frac{\partial m_{exh}}{\partial t} + \frac{\partial m_{fuel}}{\partial t} \tag{3.18}$$

The mass in the cylinder is changing due to the mass flows ($\dot{m}_{int}$ and $\dot{m}_{exh}$) of the gas exchange through the intake and exhaust valves and the injected fuel m_{fuel}. According to the first law of thermodynamics, the conservation of energy can be expressed as

$$\frac{\partial U}{\partial t} = -p_{cyl} \frac{\partial V_{cyl}}{\partial t} + \frac{\partial Q_b}{\partial t} + \frac{\partial Q_{wall}}{\partial t} + \frac{\partial m_{int}}{\partial t} h_{int} + \frac{\partial m_{exh}}{\partial t} h_{exh} + \frac{\partial m_{fuel}}{\partial t} h_{fuel} \tag{3.19}$$

The change in the internal energy in the cylinder $\frac{\partial U}{\partial t}$ is considered due to the work done by the volume change $\frac{\partial W}{\partial t} = -p_{cyl} \frac{\partial V_{cyl}}{\partial t}$ and the power of the combustion $\frac{\partial Q_b}{\partial t}$. The last three terms describe the enthalpies conducted and dissipated by gas exchange and injection. The heat transfer exchange by the cylinder wall $\frac{\partial Q_{wall}}{\partial t}$ is calculated similar to eq. 3.8 as

$$\frac{\partial Q_{wall}}{\partial t} = \sum_z \alpha_{wall,z} A_{wall,z} \left(T_{wall,z} - T_{cyl} \right) \tag{3.20}$$

The parameter z describes the different geometrical wall sections in the cylinder consisting of cylinder head, liner, and piston in this work. The solver of the cylinder wall temperature T_{wall} to determine the heat release coefficient $\alpha_{wall,z}$ is based on the empirical Woschni correlation (Woschni [180]). The total heat release of the combustion process is defined by the lower heating value of the gasoline lhv and the injected fuel mass m_{fuel} as

$$Q_{b,tot} = m_{fuel} \cdot lhv \tag{3.21}$$

There are several approaches available describing the conversion of the air-fuel mixture in a SI engine. A common way to predict the normalized burn rate is the equation according to the Vibe approach (Merker et al. [109]):

$$\frac{Q_b(\phi)}{Q_{b,tot}} = 1 - e^{-\alpha_c y_{bd}^{\beta_c+1}} \tag{3.22}$$

with

$$y_{bd} = \frac{\phi - \phi_5}{\phi_{bd}} \tag{3.23}$$

The normalized burn rate is mainly specified by the time-based parameters of the combustion and the form factor β_c. The start of combustion ϕ_5 is typically the moment, when 5% of the present fuel amount is burned. The burning duration ϕ_{bd} defines the period between 5% and 90% fuel conversion. The conversion coefficient α_c can be set to $\alpha_c = 6.9$ for a complete combustion. Usually, the burn rate is declared as (Merker et al. [109]):

$$\frac{\partial Q_b}{\partial \phi} = Q_{b,tot} \alpha_c \left(\beta_c + 1\right) y_{bd}^{\beta_c} e^{-\alpha_c y_{bd}^{\beta_c+1}} \tag{3.24}$$

However, it is complex to investigate the effect of various calibration strategies to the corresponding engine operation regarding fuel consumption and emissions due to this empirical approach above. For example, variable ignition timings and the corresponding ignition delay are not considered. Furthermore, the determination of engine-out emissions requires more detailed information about the combustion process and conditions in the cylinder.

As a central issue of the presented methodology in this work, the SI combustion process is predicted by an entrainment model (Nefischer et al. [119]) which is updated with a newly developed quasi-dimensional turbulence sub-model (Grasreiner [48], Grasreiner et al. [49]). In figure 3.8, its structure including the important modules of ignition delay, flame surface, laminar and turbulent flame speed, and turbulence model is illustrated schematically.

The basic idea of the phenomenological model is adapted from the approach dividing the combustion chamber into three elementary zones during flame propagation, as shown in figure 3.9 (Grasreiner [48]):

- The unburned zone "u" containing the homogenous fresh air-fuel mixture.

- The burned zone "b" including the products of a complete reaction.

- The entrained zone "e" consisting of the burned masses and an entrained, but not yet burned ring "e-b" described by intermediate stages of the global one-step reaction.

Due to this determination of different zones, the instantaneous heat release rates, the density step within the flame front and the compression of unburned mixture can be calculated by the entrainment model (Tallio and Colella [162]).

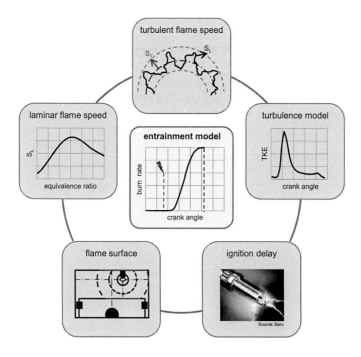

Figure 3.8.: Modules of the entrainment model

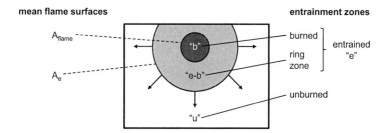

Figure 3.9.: Multi-zone division for thermodynamics in the entrainment model according to Grasreiner [48]

The fundamental calculations are introduced in principle in the following (more details are presented in Nefischer [118]). The change of mass that is caught by the flame can be expressed by the entrainment equation according to Blizard and Keck [7]:

$$\frac{\partial m_e}{\partial t} = \rho_u A_e u_e \tag{3.25}$$

49

where ρ_u is the density of the unburned gas mixture and the entrained surface is considered by A_e. The velocity of the flame front u_e describes the entering of unburned eddies into the reaction zone. Several approaches to determine the entrainment velocity can be found in the literature [3, 109, 118].

$$u_e = f\left(s_L, s_T\right) \tag{3.26}$$

Detailed information about the determination of the entrainment (u_e) and turbulent (s_T) flame speed and the consideration of the ignition delay are mentioned in Grasreiner [48]. The laminar burning velocity (s_L) can be expressed by an approach of Metghalchi and Keck [110] depending on the fuel, stoichiometric air-fuel ratio λ, residual gas content Y_{EGR}, cylinder pressure p_{cyl}, and temperature of the educts T_u in the unburned zone:

$$s_L = s_{L,0} \left(\frac{T_u}{T_o}\right)^{\alpha} \left(\frac{p_{cyl}}{p_0}\right)^{\beta} \left(1 - 2,06 Y_{EGR}^{0,77}\right) \tag{3.27}$$

The laminar burning rate $s_{L,0}$ at standard conditions (T_0, p_0) is scaled to the actual conditions in the unburned gas mixture (T_u, p_{cyl}) and can be calibrated to various fuels (e.g. propane, methanol, and isooctane) by the exponents α and β. The influence of the rate of residual gas is considered by the last term including the parameter Y_{EGR}.

The burn rate of the combustion can be defined as

$$\frac{dm_b}{dt} = \frac{m_e - m_b}{\tau} \tag{3.28}$$

with the characteristic Taylor microscale τ by Tennekes and Lumley [163]

$$\tau = \sqrt{15 \frac{\nu l_T}{v'} \frac{1}{s_L}} \tag{3.29}$$

As obvious from equation 3.28, the determination of the burn rate is influenced directly by the mass contained in the entrained zone. Last one depends on the flame propagation and thus, on the reaction surface A_e. Since local deformations of the flame front cannot be predicted due to the quasi-dimensional modeling approach, a mean entrainment surface is calculated assuming a spherical shape of the flame. This implementation is adopted from Blizard and Keck [7] and Morel et al. [117] incorporating following key properties:

- perfectly spherical shape of the flame front during propagation.

- center of the flame at a fixed location independent from the real ignition source of the spark plug.

- simplified geometry with pancake shaped combustion chamber and flat piston crown.

- geometrical limitation of flame propagation by defining wall contacting areas "inactive" for further entrainment.

In figure 3.10, the modeling of the flame surface is presented for a specific time step, exemplarily. According to this approach, the entrainment surface A_e can be defined as function of the constant flame center s_{SP}, flame radius r_F, radius of each spherical segment r_S, piston position s_{pist}, and fixed combustion chamber geometry (e.g. radius of the cylinder bore r_{cyl}):

$$A_e = f\left(r_F, s_{pist}, |\vec{r}_S|\right) \tag{3.30}$$

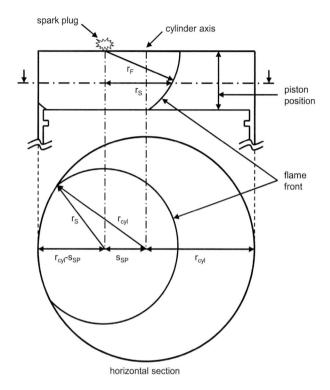

Figure 3.10.: Determination of the flame surface according to Grasreiner [48]

The entrained volume is determined by summing up the overall surface due to cumulation of the discretized sphere partitions (horizontal slices). The wall contacting areas of the flame are subtracted from the spherical effective surface slowing the entrainment. Additionally, the effect of the quenching distance near the walls is included (important for the emissions prediction as presented in section 3.6.2). Due to an additional ignition delay sub-model (Grasreiner et al. [50]) and the calculation of the flame propagation in crank angle based time steps, variable ignition timings and the resulting flame-wall-interaction (cylinder head, cylinder liner and top piston areas) can be considered. This can be an important part in the analysis of various engine calibrations.

Significant influence on the prediction of the burn out phase is achieved by the definition of the flame center initialization. It does not necessarily correlate with the real location of the spark plug. The effect of an eccentrically shifted center of the spherical flame shape away from the cylinders vertical axis is investigated in Grasreiner [48] and Grill [51]. Further information about the modeling of the flame surface is described in the literature [7, 117, 118].

According to equation 3.20, the temperatures of the cylinder walls effect the heat release including internal engine cooling. Especially during the determination of fuel consumption in driving cycles, the impact of the engine warm-up has to be considered. In addition, friction on engine system level depends mainly on temperature and speed of operation (Merker et al. [109]). Therefore, temperatures of cooling water and oil are defined from the corresponding measurement of the particular driving cycle. This simplification is necessary, since a detailed thermal simulation of engine heating effects would enlarge the transient engine model considerably and computing time would increase tremendously. For this purpose, the friction modeling is based on maps providing a negative torque for a defined combination of cooling water temperature and engine speed. The input data is provided by results of stationary simulations based on a 1D engine model including detailed thermal circuits offering engine maps considering the whole range of operation points. The process of generating these friction maps is illustrated schematically in figure 3.11 and described in the following.

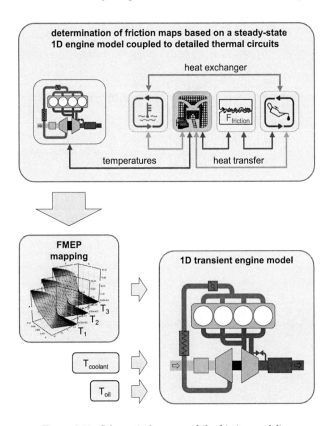

Figure 3.11.: Schematical process of the friction modeling

In order to determine the friction maps, the calculation of gas exchange and combustion is coupled to sub-models describing the thermal flow in the engine (Seider and Bet [152]). The simulation method of this thermal interchange is presented simplified by the thermal 1D model in the upper half of figure 3.11. Based on the temperatures (ruby colored) determined in the cylinder and the defined heat transfer coefficients, the thermal conduction from the cylinder walls to the engine oil and cooling water is calculated in a structural model. Especially, the characteristics of the structure in the cylinder head is modeled adequately. The heat exchanges (orange lines) are predicted by additional sub-models of both fluid circuits, oil and cooling water (Merker et al. [109]). Thereby, the inlet temperatures of the fluids are set to constant values in the heat exchanger (blue line) representing specific operation conditions of the engine. According to the temperature-conditioned operation, the corresponding friction torque can be correlated to the engine map.

Exhaust system

The separated flows of the exhaust gases from each cylinder are channeled from the exhaust manifold to the twin scroll turbo-charger. The modeling approach of the turbine is similar to the compressor described above. Turbine and compressor are coupled by balancing their power dependent on mass flow and difference of enthalpy (Roithmeier [140]). Integrated in the object of the turbo-charger, the wastegate is modeled as bypass of the main flow passing around the turbine. At the real engine, this wastegate consists of a flap gate actuated by an electric motor. However, an orifice connection with variable cross-sectional area controlled by adjustable discharged coefficients is used in this work, comparable to the modeling approach of the throttle. Here, the correlation between the actuation signal of the ECU and the flow characteristics is not implemented as profile versus opening angle, since no adequate measurements are available. Instead, the flow rate is adapted by a proportional-integral-derivative (PID) controller with respect to the requested pressure in front of the throttle (boost pressure). The modeling of the remaining exhaust system is strongly simplified to optimize computing performance, for example. The catalyst consists of a defined volume with predetermined pressure drop and changing in temperature. Sound absorbers are combined to one flow restrictor. At the end of the exhaust system, the same environment conditions as at the intake are specified.

In summary, the virtual engine system includes a crank-angle based calculation in combination with a phenomenological combustion model according to the requirements of chapter 2.2.1. Considering the geometrical data of the real engine, the air volumes can be modeled with high level of detail. This results in a good matching of its inherent gas dynamics regarding air flow, pressures, temperatures, and heat transfers. However, a moderate computing time to enable the simulation of entire driving cycles is still provided.

3.3.2 *Model calibration*

The parameterization of the components in the engine model is mainly based on geometrical and material properties. The geometries like pipe length and diameter, cylinder dimensions, and valve lift characteristics are adapted directly from the real engine. The coefficients of wall friction and heat transfer depend on the material database. Additional information belongs to specific test benches of individual components like measurements of the turbocharger and discharge coefficients of the valves. However, all these parameters can be specified by measurable characteristics of the engine. More difficult is the calibration of the entrainment model to match the combustion performance. This is achieved with the help of a pressure trace analysis (PTA) implying measurements from the engine test bench. Therefore, only one cylinder with its connections to intake/exhaust manifold is extracted from the whole engine model. Additionally to the gradient of the cylinder pressure, the measured pressure profiles at sensor positions in the inlet and exhaust manifold are applied as boundary conditions in the simulation. The air mass flow is matched to the measured one by adjusting the absolute value of the intake pressure. Based on the measured cylinder pressure, burn rates of the combustion can be predicted by the evaluation of the first law of thermodynamics (eq. 3.19).

The variable parameters of the entrainment model are identified in a subsequent simulation by calibration. For this purpose, results are matched either to the characteristics of the measured cylinder pressure or to the determined burn rate from the PTA. This iterative process is implemented into an automated optimization introduced by Grasreiner [48]. Here, five operation points covering the interesting range of the engine map are chosen. Following working steps are repeated in an automated workflow until the cumulated error values reach the required quality (Knoll et al. [83]):

1. defining a new set of parameters (optimization routine)

2. simulating steady-state the working cycle including combustion (engine model)

3. evaluating the error due to the difference of cylinder pressure (post processor)

4. comparing error criterion with accepted limits (optimization routine)

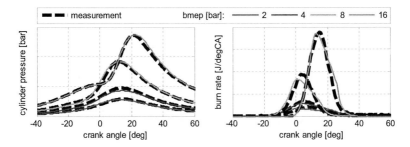

Figure 3.12.: Matching of cylinder pressure and burn rate in steady-state operation at engine speed 2000 min^{-1} and several loads (brake mean effective pressure (bmep))

Exemplarily, some results of the calibration process are shown in figure 3.12. It exhibits a good matching quality in several steady-state operating points regarding cylinder pressure and burn rate (measured reference obtained from the PTA). Indeed, similar confirmations are obtained in the entire steady-state operating range (different engine speeds and loads as presented in Nefischer et al. [118, 119]).

3.4 VEHICLE MODEL

A sub-model of the vehicle powertrain is implemented to enable the transient analyses of fuel consumption efficiency and emissions regarding government regulations. With the help of a mechanical model, the instantaneous torque determined by the engine simulation is converted into a translational, longitudinal motion of the vehicle. For the purpose of this study, a vehicle with front-wheel drive and a manual gear transmission is considered (specified in table 4.1).

The powertrain of the vehicle is basically modeled in a rigid 1D-multibody model except of one rotational elastic element. It is integrated in the same simulation tool as used for the engine calculations. The structure of the vehicle model including the connections between the components is illustrated schematically in figure 3.13. The powertrain is connected mechani-

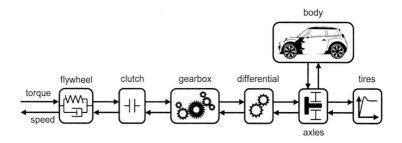

Figure 3.13.: Schematic structure of the vehicle model

cally to the cranktrain of the combustion engine by a rigid shaft. The only flexibility inside of the powertrain model is a virtual flywheel modeled as spring-damper element avoiding the transformation of torque peaks from the thermodynamic engine model to the powertrain or vehicle. Only the torsional stiffness is defined, while the torsional damping rate is unconsidered in this work. After the flywheel, a clutch element controls the transfer of the torque into the gearbox. The actuation is implemented by a parameter transmitted from the driver model. In case of start moving the vehicle at zero velocity, an additional control loop for the clutch is applied. It includes a characteristic curve defining a relation between clutch position and engine speed (figure 3.14) to avoid stalling the engine during the closing of the clutch. If engine speed drops to idle speed (table 4.1) or beneath, the clutch is opened again. The

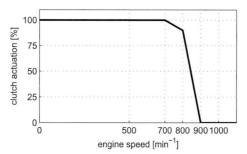

Figure 3.14.: Characteristic line to control the clutch in drive away phases with vehicle start speed = 0km/h

engine speed is rising due to the reduced resistance, followed by a reversal in the clutch actuation back to closing. This feed-forward control reproduces the natural actuating behavior by the real driver and enables a smoothly start moving of the vehicle. When the clutch is fully closed, the torque from the combustion engine is transferred lossless to the gearbox.

Gearbox and differential are mechanical components connected to their interfaces by an input and output shaft. They are modeled as kinematic transmission ratios with corresponding efficiency maps to look up the power losses due to friction at its operation point. For each gear and the final drive ratio, efficiency maps (η) as functions of shaft torque and speed of rotation are defined. The transient behavior is considered by the gear dependent mass moment of inertia. In the element of the gearbox, the kinematic ratios for each gear are specified separately. The rotating speed ratio from the input to the output shaft (referred by the subscripts in and out) is determined by kinematic constraints as

$$\omega_{out} = \frac{\omega_{in}}{ratio_{gear}} \qquad (3.31)$$

where ω represents the revolution speed and $ratio_{gear}$ the transmission ratio. The power losses imply a torque (TQ) reduction at the output shaft (Millo et al. [112]):

$$TQ_{out} = \eta \cdot ratio_{gear} \cdot TQ_{in} \qquad (3.32)$$

In this simulation methodology, the current gear selection and the moments of shifting during the entire driving cycle are provided according to the corresponding measurement or specification. During gear shifting, the vehicle and the engine speeds have to be decoupled kinematically by the clutch. The actuation is controlled by the clutch parameter from the driver model, as mentioned in section 3.2. The delay of the shifting process is defined by a fixed time period. A differentiation between the gears is possible in this model, but not applied in this work.

Coming from the differential, the driving torque is transmitted to the front axles and hence, to the wheels. The wheel suspension also connects the powertrain to the vehicle body, but there are no vertical spring and damper elements modeled. In this work, vertical vibrations of the body for e.g. driving comfort are not considered. Usually, driving cycles for fuel consumption analysis do not affect limit ranges of high driving dynamics like e.g. racing conditions. Thus, the movement of the vehicle body is not required to calculate accurate acceleration and braking behavior.

The transfer of wheel torque to the road surface is achieved by tire elements which include a slip model as well as a function of the rolling resistance of the tires. Since the vehicle dynamic is limited to longitudinal motion, detailed models dealing with lateral behavior can be neglected. If required, user defined equations, e.g. based on the common "Magic Formula" tire model according to Pacejka [128], can be adapted. The deformation of the tire is simplified by specifying the rolling radius of the loaded tire as a function of tire speed. The torque applied to the wheel axle is transformed into a traction force considering a rolling motion including slip between the tire and road surface. According to this approach, the maximum force generated at the road interface in longitudinal direction $F_{x,gross}$ can be assumed as proportional function of the vertical load on the wheel F_{load} and the friction coefficient for roads $\mu_{friction}$.

$$F_{x,gross} = \mu_{friction} \cdot F_{load} \qquad (3.33)$$

The tire friction is determined by a map of the coefficient μ as a function of vehicle speed v_{veh} and tire-vehicle slip λ_{tire}. Last one is defined as

$$\lambda_{tire} = \begin{cases} 0 & v_{veh} = 0, \omega_{tire} = 0 \\ \frac{\omega_{tire}r_{tire}-v_{veh}}{|\omega_{tire}r_{tire}|} & |\omega_{tire}r_{tire}| > |v_{veh}| \\ \frac{\omega_{tire}r_{tire}-v_{veh}}{|v_{veh}|} & v_{veh} \neq 0, |v_{veh}| \geq \omega_{tire}r_{tire} \end{cases} \tag{3.34}$$

where ω_{tire} is the tire speed and r_{tire} is the tire rolling radius. The net tire tractive force, applied to the vehicle, is

$$F_x = F_{x,gross} - F_{roll} \tag{3.35}$$

where F_{roll} is the rolling resistance force depending on the normal load F_{load}:

$$F_{roll} = c_{roll} \cdot F_{load} \cdot \cos\alpha \tag{3.36}$$

In this model, the tire rolling resistance coefficient c_{roll} is defined as function of tire speed ω_{tire}. The slope of the road is considered by the angle α. The wheel torque transferred to the road is limited by

$$-F_x \leq \frac{TQ_{wheel}}{r_{tire}} \leq F_x \tag{3.37}$$

The brakes are modeled as an additional torque loss reducing the net torque applied to the wheel axle:

$$TQ_{wheel} = TQ_{drive\_shaft} - TQ_{brake} \tag{3.38}$$

The brake torque TQ_{brake} is determined as a function of wheel axle speed and brake pedal position. Last parameter is controlled and communicated by the driver actuation model.

In the element of the vehicle body, several parameters specifying the individual characteristics of each car type can be configured. For example, the vehicle mass m_{veh} is specified here and additional passenger and cargo mass can be defined as well. Geometry data of the chassis like vehicle wheel base and the location of the mass center in horizontal and vertical relation to the axes are required to determine the normal loads of each tire. Additional to the rolling resistance further losses of the vehicle have to be considered:

Aerodynamic resistance: According to the vehicle parameters of the frontal area A_{f_veh} and the aerodynamic drag coefficient c_{drag}, the aerodynamic force can be calculated as

$$F_{aero} = \frac{1}{2} \cdot \rho_{air} \cdot A_{f\_veh} \cdot c_{drag} \cdot v_{veh}^2 \tag{3.39}$$

where ρ_{air} is the air density.

Acceleration resistance: Depending on the vehicle longitudinal acceleration ($\dot{v}_{veh}$), the mass of the vehicle m_{veh}, and inertial resistances of each component of the drive-train $J_{components}$, the acceleration force can be determined:

$$F_{acc} = f\left(m_{veh}, J_{components}, \dot{v}_{veh}\right) \tag{3.40}$$

Slope resistance: The road slope can be considered as

$$F_{slope} = m_{veh} \cdot g \cdot \sin\alpha \tag{3.41}$$

where g is the gravity acceleration and α is the road slope. The slope resistance can be neglected in this work, since in driving cycles regulated by law to determine fuel consumption and emissions only flat roads are considered.

Once all resistances are defined, the vehicle equilibrium equation can be calculated as

$$F_{acc} = \sum_{tire\ i=1}^{4} F_{x,tire\ i} - F_{roll} - F_{aero} - F_{slope} \tag{3.42}$$

$F_{x,tire\ i}$ represents the longitudinal traction and braking force generated by each tire $tire\ i$.

In summary, the vehicle speed is determined by considering the powertrain kinematics and power losses are calculated due to the modeling of the corresponding powertrain characteristics according to the requests of chapter 2.2.2. The specifications to define the kinematics and kinetics are adopted from module test benches or computer-aided design (CAD) models. The information about the driving resistance of the vehicle are transferred from coast down tests that are already performed for the measurements of fuel consumption on a roller chassis dynamometer. Additionally, the transient simulation is not restrained regarding computing time due to the relative basic modeling approach of the vehicle.

3.5 COUPLED MODEL

After modeling the single part models, the interfaces are connected to one global model including simultaneous processing. But not only the time synchronization, also the communication between the models itself has to be ensured (see requirements in chapter 2.3). The coupling implicates more complexity into the system. Discrepancies of one sub-model can affect important input parameters of another calculation causing strong deviations of the overall simulation results.

In this simulation methodology, both systems of ECU/driver control and engine/vehicle are modeled in two different simulation tools. For the main time period of the ECU model, a range of 10 milliseconds is defined. This discrete time constant is not useful for the simulation of gas exchange and combustion, since their effects are heavily dependent from engine speed and require more detailed time periods. For this reason, the engine model has to keep its crank train synchronized calculation. This requires a synchronization in time at the interchange of parameters between both simulation environments. The interconnection is performed by predefined elements that already exist in both tools and are provided by the manufactures by default. In this work, the coupled simulation is fully controlled by the ECU model operating as master. Consequently, the slave function is fulfilled by the engine/vehicle model integrated with its own solver. The engine computation is invoked by the ECU model and the exchange of parameters is controlled by the linking objects in both simulation environments. Since the interface is defined as standardized configuration, sub-models can be exchanged easily enabling the investigation of various combinations of engine control and vehicle/engine types. Nevertheless, reasonable interval timing for the communication has to be chosen (ideally the time period of the ECU model) and the synchronization in time has to be proven.

Due to physical restrictions of the real engine, the actuation parameters cannot be changed within a working cycle. This is observed by filters in the engine model. For each cylinder, the signals are kept constant for defined time durations within its actual working cycle. This important effect is exemplified for one cylinder in figure 3.15 showing a short section of the acceleration at around 800s without gear shifts during the NEDC. While the ECU is changing constantly the valve lift at the intake (LIFT) and ignition angle (IGN) to achieve the rising power demand (demonstrated by the profile of the torque at the engine shaft in the upper plot), the actuation parameters in the engine model are set to constant values during specific time periods. Last ones are defined as the opening time of the intake valves for the valve lift and the combustion duration (burn rate) for the spark timing. The same principle is applied to the parameters of valve timing and air-fuel equivalence ratio.

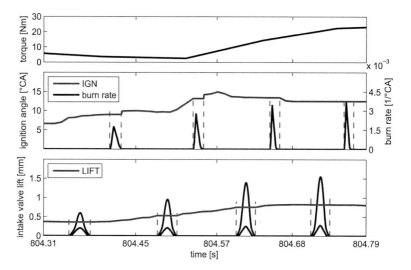

Figure 3.15.: Signal filtering of the actuation parameters within the engine model exemplified for one cylinder

The second important requirement of chapter 2.3 is the correct communication in terms of an accurate transfer of calculated parameters between the models itself. Rather simple is the interconnection of the sub-models integrated in the same simulation environment. For example, the virtual vehicle is connected to the combustion engine model by only one mechanical link from the cranktrain to the primary side of the clutch exchanging engine speed and torque. Another simplification is the straight implementation of the driver control into the ECU model.

More extensive is the transfer of the actuator and sensed parameters between control system and the engine/vehicle model (compare figure 2.7). On the feed-forward side, the actuation of the engine components is reduced to the main parts effecting gas exchange and combustion. The vehicle sub-model is actuated only by the driver control due to exchange of the signals clutch position, gear number, and actual vehicle speed. Sensed quantities of actual states from engine and vehicle are provided on the feed-back side. The kind of signal is based on the real sensors of the vehicle. Most parameters coming from the engine model are averaged over each combustion cycle, since no sensors with higher time rates generally can be found in a series-production vehicle. Attention has to be paid to consistent units used for each individual parameter.

Finally, this yields a closed-loop control system nearly without any impacts from outside of the coupled simulation environment. Only a few signals are still required from measurements for a precise validation of the simulation results, listed in table 3.1. The parameters with regard to the particular driving cycle like vehicle speed, gear number, and clutch actuation are implemented as time dependent profiles directly from measurements. The profile of the

Required input	Implementation
demanded vehicle speed	profile as function of time
gear number	profile as function of time
clutch actuation	profile as function of time
air-fuel ratio	profile as function of time
oil temperature	profile as function of time
cooling water temperature	profile as function of time
torque loss by auxiliaries	profile as function of time
intake air temperature	profile as function of time
ambient conditions	constant values

Table 3.1.: Summarizing of required input data for the overall coupled simulation.

air-fuel ratio is also adopted from the corresponding measurement to eliminate its effect for the prediction of fuel consumption and emissions in this simulation methodology. Investigations about its sensitivity and a modeling approach to create a virtual signal are presented in chapter 5.2. The engine temperature has an important impact on the fuel efficiency and emission emitting. As described in section 3.3.1, a detailed thermal simulation considering engine structure effects is required to provide the calculation concerning oil and cooling water temperature. Such a model extension is used for the determination of the friction map handling only steady-state engine operation, but it is too time-consuming for transient simulations. In this work, temperature profiles as function of time from measurements are provided to reproduce an accurate heating-up behavior of the engine. Another essential parameter affecting the precise validation of vehicle simulations in driving cycles is the torque loss of the auxiliaries. Since electrical consumers are not considered in the vehicle model, the measured signal is included in the engine calculations. In order to avoid the influence of deviations in the ambient conditions and the intake air temperature as identified in section 3.3.1, the measured pressure and temperature are defined as constant values at the engine boundary objects and the intercooler is conditioned by the corresponding temperature profile of T_{air_int}. All inputs from measurements are imported and controlled by the ECU model. Thus, these parameters have to be implemented additionally at the interchange of the coupled simulation environment.

3.6 IN-CYLINDER EMISSIONS

The presented simulation methodology is not only built to predict fuel consumption in driving cycles, but also to estimate the gaseous in-cylinder emissions HC, NO_x, and CO. These are important for the dimensioning of the exhaust aftertreatment components. Using the gained information about the combustion process in the cylinder from the engine model, in-cylinder emissions are calculated in separated models added to this simulation methodology as post-processing part. This process is splitted into different models for HC and NO_x/CO due to their individual formations mechanisms. While the HC-model uses physical gaseous effects, the NOx/CO-model is based on chemical kinetic reactions.

3.6.1 NO$_x$-and CO-model

Detailed information of each working cycle can be provided for the proposed engine-out emissions sub-model due to the crank-angle based engine simulation in this work. As required in chapter 2.4, the formation of the in-cylinder emissions NO$_x$ and CO is described by a chemical reaction kinetics model. In figure 3.16, the calculation process of the emissions evaluation is illustrated schematically. On the basis of the engine and entrainment model, the combustion profile in combination with dependent parameters like e.g. pressure and temperatures in the cylinder are determined. Implemented as post-processing tool to the gas-exchange simulation, the characteristics of temperatures and gas compositions are converted into the concentrations of NO$_x$/CO in the burned gas by a chemical kinetic model. It consists of a 0D cylinder model including detailed chemical mechanisms adopted from Zschutschke et al. [184]. While the oxidation of fuel is defined by an iso-octane mechanism, a sub-mechanism is added to calculate the formation of NO$_x$.

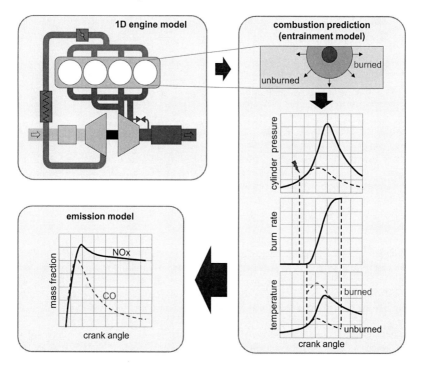

Figure 3.16.: Schematic layout of the calculation process determining in-cylinder emissions

The detailed chemical reaction equations describing the combustion process are implemented in a model of a multi-zone reactor. It is based on a chemical solver similar to Linse [99] computing the conservation equations of mass, species, and energy of every sub-volume of the reactor chamber. In this work, the combustion chamber is divided into two zones,

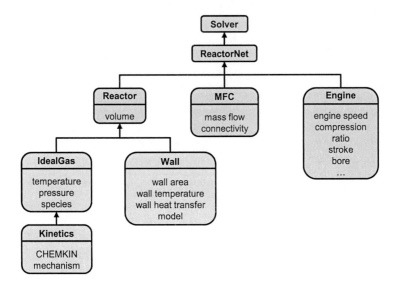

Figure 3.17.: Schematic layout of the top-level sturcture of *ProEngine* [99]

unburned and burned gas mixture. However, boundary conditions like e.g. injection and heat losses are considered, as well. The detailed reaction kinetics are realized in *ProEngine*, a tool for reacting-flow simulations including a 0D chemistry solver to evaluate properties and chemical source terms appearing in the governing equations. It enables object-oriented programming to combine various types of reactors, reacting mixtures, and ODE integrators. The top-level structure of *ProEngine* and its object types are illustrated in figure 3.17 schematically. Following various types of objects can be applied in the *ProEngine* code (according to Linse [99]):

Kinetics: The reaction mechanism is implemented in the *Kinetics* object to calculate the reaction rates. There is no limitation in the number of both, species and reaction equations.

IdealGas: This type computes the chemically reaction of ideal gas mixture using the equation of state and an associated kinetic object.

Reactor: This object represents a perfectly-stirred reactor model. A network of user-defined dimension can be generated by connecting several objects of this type. Each reactor must be linked to a corresponding *Idealgas* object representing the containing fluid.

Wall: Boundary conditions by any number of walls can be added to the *Reactor* object defining its wall area, temperature, and a wall heat transfer model. By default, all walls are adiabatic.

MFC: If a network of several reactors exists, the connectivity and massflow between each of them can be defined by the *MFC* object.

ReactorNet: The coupling of the *Reactor* objects is controlled by this type providing a data structure for a possible set of reactors and information about the way of interchange.

Engine: Engine specific data like engine speed, compression ratio, bore, and stroke etc. is applied by the *Engine* object including predefined methods to determine the piston movement and the corresponding change in cylinder volume and surface area.

Solver: This type contains an ODE integrator and a DASSL solver (see Petzold [131]) to solve the differential equation $f(t, y, y') = 0$ in all reactors of the model.

In the present approach, the unburned and burned gas mixture is described by two perfectly-stirred reactors. The specification like e.g. composition and volume can be adopted directly from the entrainment model. The state of each reactor model is defined by its mass, species composition, and temperature. Pressure is assumed to be constant among both zones. Between these two reactors, gas masses, species, and energy can be interchanged. The individually initialization of both objects and the massflow rates between them are input parameters from the combustion computation of the engine model.

According to Zschutschke et al. [184], the chemical mechanisms are differentiated between an iso-octane mechanism and a sub-mechanism for the formation of NO_x in-cylinder emissions. First one describes the oxidation of hydrogen from the fuel (here iso-octane is estimated) to the major products, carbon dioxide and water. Here, the RedMech, a slightly modified version originally from Golovitchev (Golovitchev [45], Huang et al. [65]), is chosen. The combustion products are calculated without the NO_x determination in a chemical equilibrium solver using 398 reaction equations considering 84 species. This postulates a very fast conversion of the unburned gas mixture compared to the NO_x formation process. The NO_x in-cylinder emissions are then computed by the same two-zone model using an adapted sub-mechanism from Mauss (Ahmed et al. [1]) including three species and eight reaction equations.

The chemical kinetics model does not interact with the combustion model, since the formation process in the post-flame region does not have an impact on the flame progress and the overall heat release. The temperature in the burned zone and the consumption rate are solely provided by the entrainment model. The approach of this model using detailed chemical kinetics considers only the thermal formation of NO_x. Since the entire flame structure needs to be resolved to determine the hydrocarbon radicals, mechanisms of prompt NO_x are not simulated, for example. Even, there is a separate calculation of both, NO and NO_2, only their summation counts in the matching with measurements. Generally, the determination of further NO_x variations can be ignored, since NO plays the essential role (v. Basshuysen and Schäfer [170]). The concentration of CO in the burned gas follows inevitably from the iso-octane mechanism applied in combination with the sub-mechanism to identify the formation of NO_x in-cylinder emissions.

Results of steady-state engine operation

The matching of the emission model is achieved by steady-state operating points using measurements from an engine test bench. In figure 3.18, the simulated results of the entire engine map are compared to the corresponding measured amounts per working cycle. The diagonal

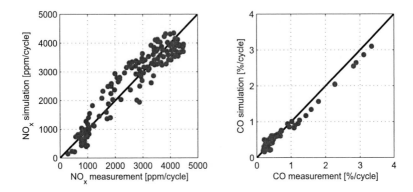

Figure 3.18.: Matching the NO_x and CO engine-out emissions of the steady-state engine map simulation

lines present the perfectly matching between both values of simulation and measurement. In the left plot, the NO_x emissions spread in a range of 1000ppm around the congruence line. Sufficient quality is achieved at low values representing an important part in legal regulated driving cycles. Only small discrepancies can be seen at the validation of the CO emissions on the right side. Here, an additional adjustment is implemented in the evaluation of the simulated results. Considering the residual gas content Y_{EGR}, an empirical offset is added to the primary calculated amount CO_{model} as

$$CO_{adj} = CO_{model} + 0.6\% \cdot \left(\frac{35\% - Y_{EGR}}{35\%} \right) \tag{3.43}$$

As result of the adjustment, low amounts until 1% range around the 100% matching, while a constant underestimating still occurs at several higher values.

3.6.2 Unburned hydrocarbon model

In the calculation process, the crank-angle based engine simulation provides pressure and temperature profiles of each working cycle as input information of the proposed engine-out HC sub-model. It incorporates the key mechanisms in the formation of HC in-cylinder emissions according to 2.4.3. The calculation process is illustrated schematically in figure 3.19. At warm engine condition, it includes the main sources piston crevice and oil layer. In order to reproduce the increased HC emissions during engine warm-up, the mechanism of wall quenching is added by a function dependent on engine temperature T_{eng} (represented by the coolant):

$$f_{quenching} = c_{quenching} \left(T_{eng,warm} - T_{eng} \right). \tag{3.44}$$

After summing the results of the different formation processes, the amount of HC is reduced by in-cylinder oxidation.

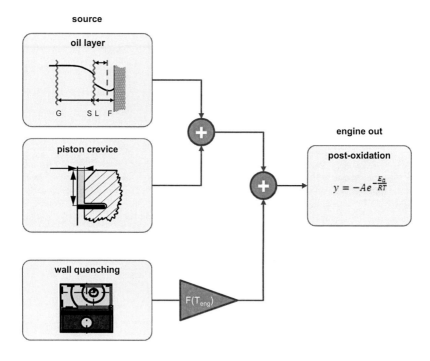

Figure 3.19.: Schematic layout of the calculation process determining in-cylinder HC emissions

Piston crevice

The largest source of engine-out HC emissions is the crevice volume between liner, piston, and top piston ring. The storage of unburned gas mixture and the corresponding flow into and out of the piston crevice is determined using the equation of state for ideal gases as

$$p_{cyl}V_{cr} = m_{HC,cr}RT_{cr} \qquad (3.45)$$

with the specific gas constant R from the gas exchange simulation. The mass of HC in the crevice $m_{HC,cr}$ is calculated by starting with an initial mass at start of combustion (SOC) and integrating its change till end of combustion (EOC) with respect to crank-angle φ (second term):

$$m_{HC,cr} = \underbrace{\frac{V_{cr}p_{cyl}}{RT_{cr}}\bigg|_{SOC}}_{\text{initial value}} + \underbrace{V_{cr}\int_{SOC}^{EOC}\frac{1}{RT_{cr}}\frac{\partial p_{cyl}}{\partial\varphi}d\varphi}_{\text{crank-angle based term}} \qquad (3.46)$$

It is assumed that in the whole combustion chamber (including piston crevice) the pressure is equal to the cylinder pressure p_{cyl}. The temperature in the crevice volume T_{cr} can be calculated in different ways. Here, the arithmetic mean of the wall temperature T_{wall} and unburned gas temperature T_u is assumed (Merker et al. [109]):

$$T_{cr} = \frac{T_{wall} + T_u}{2} \tag{3.47}$$

The volume of the piston crevice V_{cr} is defined by bore and piston diameter d and distance between piston crown and top compression ring ($s_{cr,height}$):

$$V_{cr} = \frac{\pi}{4} \left(d_{bore}^2 - d_{piston}^2 \right) s_{cr,height} \tag{3.48}$$

The flow passing the piston rings to the crankcase (blowby) can be neglected due to its small change in mass. The mass trapped between top and second ring is not accounted for, since this is at most about 20% of the maximum mass in the volume above [12]. In some studies [164, 178], the location of the spark plug seems to affect the amount of HC emissions. The investigated engine in this work has an almost central spark position and therefore, influences of the location of the flame front with respect to the piston crevice are not considered.

Near EOC, the HC mass flow into the combustion chamber is determined due to the mass change in the crevice. It is assumed that gas mixture leaving the crevice during combustion is still burned completely. In figure 3.20, the mass flow and the resulting mass of HC emissions (normalized to its maximum value) remaining in the cylinder after combustion are exemplified for steady-state warmed-up engine operation.

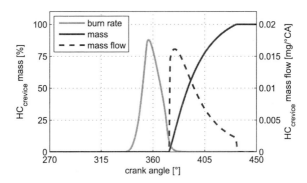

Figure 3.20.: Relative HC mass and mass flow from the crevice in the combustion chamber at warmed-up engine operation of 1000min^{-1} and 2.5bar *bmep*

Oil layer

The influence of the oil layer to the engine-out HC emissions is described in many studies [98, 150, 164, 182]. All of them are based on the assumption of equilibrium absorption of the vaporized fuel into the oil layer considering mass diffusion resistance in the liquid (oil film) and the gas phases. The change in concentration of dissolved fuel in the oil film $\frac{\partial X}{\partial t}$ can be expressed by Fick's second law (Eqn. (3.49)) describing a mass diffusion. Hereby, the molecular diffusion coefficient $D_{fuel,oil}$ for the oil film with the thickness δ_{oil} is assumed as constant.

$$\frac{\partial X}{\partial t} = D_{fuel,oil} \frac{\partial^2 X}{\partial x^2} \qquad \forall \left[0 \le x \le \delta_{oil} \right] \tag{3.49}$$

Diffusion coefficient: Several approaches to define the diffusion coefficient of fuel into oil can be found in the literature. Most of them ([164], [66], [98], and [157]) are based on the definition by Wilke and Chang [179]:

$$D_{fuel,oil} = \left[7.4 \times 10^{-8} \frac{(\psi M_{oil} \cdot 10^3)^{0.5} T_{oil}}{(\mu_{oil} \cdot 10^3)(\vartheta_{fuel} \cdot 10^6)^{0.6}} \right] \cdot 10^{-4} \left[\frac{m^2}{s} \right] \tag{3.50}$$

Depending on the solvent, Wilke and Chang [179] recommend to choose for the association factor $\psi = 2.6$ if the solvent is water, 1.9 if it is methanol, 1.5 for ethanol, and 1.0 for all unassociated solvents. Thus, the association factor is set to unity for following investigations in this work. Frölund and Schramm [37] use the expression from [179], as well, and specify it for iso-octane in *SAE 10W-40* oil:

$$D_{fuel,oil} = \left[4.4485 \times 10^{-9} \cdot T_{oil}^{1.47} \left(10^{10^{-2.936 \log_{10}(T_{oil}+7.675)}} - 0.7 \right)^{-0.70914} \right]$$
$$\cdot 10^{-4} \left[\frac{m^2}{s} \right] \tag{3.51}$$

Another empirical approach from Norris and Hochgreb [125] results in:

$$D_{fuel,oil} = \left[1.33 \times 10^{-7} \cdot T_{oil}^{1.47} \left(\mu_{oil} \cdot 10^3 \right)^{\left(\frac{10.2}{\vartheta_{fuel} \cdot 10^6} - 0.791 \right)} \right.$$
$$\left. \times \left(\vartheta_{fuel} \cdot 10^6 \right)^{-0.71} \right] \cdot 10^{-4} \left[\frac{m^2}{s} \right] \tag{3.52}$$

The different characteristics are presented in figure 3.21. Even for different oils with corresponding varying dynamic viscosity μ_{oil}, nearly all approaches result in a similar behavior of the diffusion coefficient. Only the insertion of the parameters from Trinker et al. [164] and Janssen [70] into the expression from Wilke and Chang [179] for the *SAE 5W-30* oil show lower values over the whole temperature range.

Dynamic viscosity: The dynamic viscosity of the oil can be determined according to Watkins [174]:

$$\log_{10} \left(\log_{10} \left(\frac{\mu_{oil}}{\rho_{oil}} + C \right) \right) = A - B \cdot \log_{10}(T_{oil}) \tag{3.53}$$

$$\mu_{oil} = \rho_{oil} \left(10^{10^{(A-B \cdot \log_{10}(T_{oil}))}} - C \right) \cdot 10^{-6} \tag{3.54}$$

The identification of A, B, and C and the specification of density ρ_{oil} and molar mass M_{oil} depend on the used oil type. Several parameter sets of different oils are summarized in table 3.2. The information about the *SAE 5W-30* oil is determined for the Castrol SLX Longlife 3 by Castrol according to Janssen [70].

Molar volume of fuel: The molar volume of the fuel ϑ_{fuel} can be determined approximately by

$$\vartheta_{fuel} = \frac{M_{fuel}}{\rho_{fuel}} \tag{3.55}$$

using the density ρ_{fuel} and molar mass M_{fuel} of the fuel according to Sodre and Yates [157].

Oil film thickness: The thickness of the oil layer δ_{oil} is estimated defining a constant value for the complete liner. An approximation from Norris and Hochgreb [125] is based

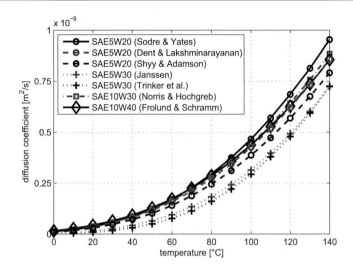

Figure 3.21.: Comparison of different diffusion coefficients of fuel into oil

Classification	A	B	C	ρ_{oil} [kg/m³]	M_{oil} kg/kmol	Source
SAE 5W-20	7.70100	3.0100	0.6	856.0	445	[157]
SAE 5W-30	7.78052	3.0085	0.8	850.0	≈495	[70]
SAE 5W-50	6.93166	2.6540	0.8	864.0	≈550	[37]
SAE 10W-30	8.17000	3.1600	0.7	-	-	[125]
SAE 10W-40	7.40552	2.8542	0.8	961.0	≈550	[37]
SAE 20W-40	8.54296	3.3015	0.7	865.0	-	[164]

Table 3.2.: Parameters of various oils

on measurements and can be calculated in respect of engine speed rpm in min^{-1} and the dynamic viscosity (considering oil temperature T_{oil} effects):

$$\delta_{oil} = \left(0.143\sqrt{\mu_{oil} \cdot rpm \cdot 60} + 2.08\right) \cdot 10^{-6} \ [m] \tag{3.56}$$

The measurements are carried out with different oils at wall temperatures of 40-85°C and engine speeds between 1000 and 2500 min^{-1}. The oil film thickness versus engine speed at several liner temperatures (assumption of $T_w = T_{oil}$) is shown in figure 3.22. The values are comparable with constant parameters in the literature of 2-5μm ([153] and [157]) and 3μm [16].

Mass transfer: The mass transfer in the absorption/desorption process is based on Henry's Law describing the interface condition between the gas and liquid phase as

$$k_H \cdot X_{component,liquid} = p \cdot X_{component,gas} \tag{3.57}$$

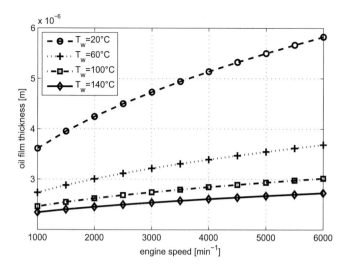

Figure 3.22.: Oil film thickness versus engine speed at several liner temperatures

with the concentrations X of the component and pressure p. The mole fractions of fuel dissolved in the oil film $X_{fuel,oil}$, and diffused in the gas mixture $X_{fuel,gas}$ can be calculated in dependency of the chemical amounts n:

$$X_{fuel,oil} = \frac{n_{fuel}}{n_{fuel} + n_{oil}} \approx \frac{n_{fuel}}{n_{oil}} \tag{3.58}$$

$$X_{fuel,gas} = \frac{n_{fuel}}{n_{fuel} + n_{air}} \approx \frac{n_{fuel}}{n_{air}} \tag{3.59}$$

A Henry Number N_H can be defined by inserting Eqn. (3.58) and Eqn. (3.59) into Eqn. (3.57):

$$k_H X_{fuel,oil} = p X_{fuel,gas}$$
$$\frac{n_{fuel}}{n_{fuel} + n_{oil}} = \frac{p}{k_H} \frac{n_{fuel}}{n_{fuel} + n_{air}}$$
$$N_H = \frac{k_H}{p} \frac{M_{oil}}{M_{air}} \tag{3.60}$$

The molar mass of air M_{air} is assumed as $28.85 \frac{\text{kg}}{\text{kmol}}$ [109].

In the literature, different definitions of the Henry's constant k_H with a strong temperature dependency for various fuel/oil combinations are mentioned. A comparison is shown in figure 3.23 and described in the following. Frölund and Schramm [37] evaluate Henry's constant for the oil layer based on measurements:

$$k_H = 10^{(7.37 \cdot \log_{10}(T_{oil}) - 16.35)} \cdot 10^{-2} \ [\text{bar}] \tag{3.61}$$

It describes the solution of iso-octane in $SAE\ 10W\text{-}30$ for a temperature range between 90-150°C. Norris and Hochgreb [125] provide an expression based on measurements of toluene in $SAE\ 10W\text{-}30$:

$$k_H = 6.8 \times 10^{-5} \cdot \left(T_{oil} - 273.15\right)^{2.25} \ [\text{bar}] \tag{3.62}$$

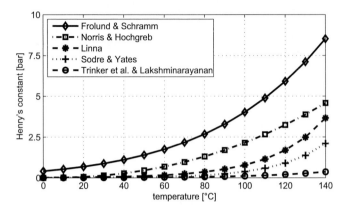

Figure 3.23.: Comparison between different determinations of Henry's constant for various fuel/oil combinations

Linna [97] used measurements of n-heptane in *SAE 5W-20*:

$$k_H = 4.0 \times 10^{-7} \cdot e^{0.0388 \cdot T_{oil}} \ [\text{bar}] \tag{3.63}$$

Henry's constant is adapted for the use of iso-octane in *SAE 5W-20* by Sodre and Yates [157]:

$$k_H = \left[e^{(0.04292 \cdot T_{oil} - 17.0045)} \right] \cdot 1.0133 \ [\text{bar}] \tag{3.64}$$

Dent and Lakshminarayanan [16] and Trinker et al. [164] define for n-octane in *SAE 5W-20*:

$$k_H = \left[10^{(-1.921 + 0.013 \cdot (T_{oil} - 300))} \right] \cdot 1.0133 \ [\text{bar}] \tag{3.65}$$

Compared to most of the mentioned studies using the differential Eqn. (3.49), the assumption of a linear concentration gradient in the oil film from Dent and Lakshminarayanan [16] is used here. However, the effects of engine speed *rpm* on the diffusion of the fuel into the oil film are accommodated by considering a penetration depth $\tau_{fuel,oil}$ expressed as:

$$\tau_{fuel,oil} = \sqrt{\frac{\pi D_{fuel,oil}}{rpm}} \tag{3.66}$$

The equation is adopted from Kreith [88] assuming a finite body to be semi-infinite in behavior for fast transients in transient heat conductions [16].

Figure 2.9 shows the concentration gradient of fuel in the different phases schematically. Between the bulk gas state (G) and the oil film (F), a gas boundary layer is assumed. The convective mass flux $\dot{m}''$ across the boundary layer of the gas phase can be defined [16] as

$$\dot{m}'' = g_{gas}^* \left(mf_G - mf_S \right) \tag{3.67}$$

with mf_G and mf_S as mass concentrations of vapor fuel diluted in the bulk gas state (G) and on the gas side of the interface (S) between both layers of the gas boundary and oil. g_{gas}^*

expresses the gas phase mass transfer conductance assuming the Reynolds analogy between heat and mass transfer which implies a Lewis Number of unity under these conditions [16]:

$$g_{gas}^* = \frac{\alpha_{cyl}}{c_p} \qquad (3.68)$$

The heat transfer coefficient α_{cyl} and the specific heat c_p are evaluated from the combustion engine model solving the correlation of Woschni [180].

In the liquid phase across the penetration depth into the oil film, the mass flux $\dot{m}''$ can be expressed as:

$$\dot{m}'' = g_{fluid}^* \left(mf_L - mf_F \right) \qquad (3.69)$$

Similarly to Eqn. (3.67), mf_L and mf_F are the mass concentrations of dissolved fuel on the liquid side of the interface (L) and in the bulk of the oil film (F). The mass transfer conductance of the liquid phase g_{fluid}^* can be written as:

$$g_{fluid}^* = \frac{\rho_{oil} D_{fuel,oil}}{\tau_{fuel,oil}} \qquad (3.70)$$

At the interface, the right hand sides of Eqn. (3.67) and Eqn. (3.69) can be equated. Using the Henry Number in Eqn. (3.60) along with algebraic re-arrangement in Eqn. (3.69), the mass flux $\dot{m}''_{fuel,oil}$ can be defined as:

$$\dot{m}''_{fuel,oil} = \frac{g_{gas}^* g_{fluid}^*}{N_H g_{gas}^* + g_{fluid}^*} \left[mf_G - mf_F N_H \right] \qquad (3.71)$$

Equation (3.71) is positive during absorption of fuel vapor by the oil film and changes sign while desorption. In order to determine the mass fraction in the bulk of the oil film mf_F, the mass of the oil film m_{oil} can be calculated from oil layer thickness, geometric data of the cylinder, and mass density of the oil. The rate of absorption/desorption is given by multiplying the mass transfer flux $\dot{m}''_{fuel,oil}$ from Eqn. (3.71) with the surface area of the oil layer in contact with the gas mixture A_{oil}. Considering piston movement by its varying position s_{stroke}, this surface area can be expressed as:

$$A_{oil} = 2 \left(\frac{d_{bore}}{2} - \delta_{oil} \right) \pi s_{stroke} \qquad (3.72)$$

The desorbed mass of fuel after combustion end is integrated until the exhaust valves open (EVO) and is an input for the in-cylinder oxidation process.

In figure 3.24, the influence of the oil temperature on the absorption process is presented for one working cycle. Thereby, the profiles of the HC mass absorbed in the oil film are referred to the maximum mass at warmed-up conditions of 100°C. Significantly more fuel is absorbed by cold oil films, but the desorption rate does not increase in the same way. Thus, the HC emissions based on the oil layer formation do not follow the maximum HC masses in the oil film stringently.

Wall quenching

The mechanism of wall quenching accounts for only a small part of engine-out HC emissions, as mentioned in section 2.4.3. In the literature, many investigations are performed about quenching with laminar or turbulent flame fronts. All the results confirm the small influence of wall quenching as source of unburned hydrocarbons [70]. For that reason, only a few

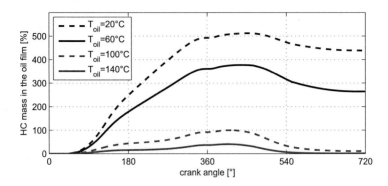

Figure 3.24.: Profiles of absorbed HC masses in the oil film for different oil temperatures at $2000\mathrm{min}^{-1}$ and 4.5bar *bmep*

models exist using the mechanism of wall quenching to calculate HC emissions. Usually, detailed chemical kinetics in combination with conservation equations of mass, species, and energy are used to determine the diffusion and oxidation of unburned hydrocarbons near the cylinder wall [70]. The physical detailed modeling requires extensive and precise information from measurements or 3D CFD simulation to specify e.g. the flow in the combustion chamber or the extinguishing of the flame front. Therefore, a modeling similar to the approach of the piston crevice mechanism is chosen.

The equation calculating the mass of unburned hydrocarbons $m_{HC,quenching}$ in the quench layer is based on the ideal gas law as

$$m_{HC,quenching} = \sum_{z=1}^{3} \left(\int_{SOC}^{EOC} \frac{p_{cyl}M_{HC}}{RT_u} \frac{\partial V_z}{\partial \varphi} d\varphi \right) \tag{3.73}$$

$$= \sum_{z=1}^{3} \left(\int_{SOC}^{EOC} \frac{p_{cyl}d_{quenching}}{(c_p - c_v)T_u} \frac{\partial A_z}{\partial \varphi} d\varphi \right) \tag{3.74}$$

with the molar mass of HC M_{HC}, specific heat for a constant volume c_v, and the quench volume V_z or surface area A_z. In order to determine the wall areas that are in contact with the flame front, the combustion chamber is divided into three zones z:

1. Cylinder head

2. Cylinder liner

3. Piston crown

In figure 3.25, the different zones are illustrated schematically. Due to the quasi-dimensional approach of the entrainment model, a spherical flame propagation is assumed [52]. Various positions of the flame front at three time steps are presented. At the beginning (a), there is no contact of the flame with the cylinder walls. Then, parts of the flame front extinguish near the wall of the cylinder head (b) and the cut surface A_{cyl_head} is >0 (dashed red line). The last snapshot (c) shows the situation near the end of combustion. Here, the flame has reached all three zones generating the red marked cut surfaces.

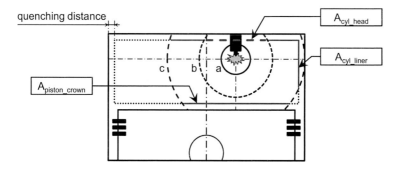

Figure 3.25.: Schematic illustration of cut surfaces at different positions of the flame front

The quenching distance $d_{quenching}$ results from a simplified empirical equation according to Peters [130]:

$$d_{quenching} = c \cdot \delta_L \tag{3.75}$$

Peters determines the parameter $c \approx$ 5-6, while the laminar flame thickness δ_L can be expressed as:

$$\delta_L = \frac{\lambda_b}{c_{p,b}s_L\rho_u} \tag{3.76}$$

The thermal conductivity λ_b and the specific heat capacity $c_{p,b}$ of the burned composition are calculated within the gas exchange simulation, as well as the density of the unburned gas mixture ρ_u. The equation of the laminar burning rate s_L (3.27) is solved in the entrainment model (presented in section 3.3.1).

In figure 3.26, the output of HC emissions due to wall quenching during the combustion is compared to the profiles of the other sources for a cold engine ($T_w = 25°C$) operating steady-state at 1000min^{-1} and 2.5bar $bmep$. The quench volume is increasing from start of combustion until the flame front covers all walls of the combustion chamber entirely. Thereafter, the HC mass of this source remains constant in the cylinder. At this time, the desorption process of the oil layer has not even started.

In-cylinder oxidation

After the end of the combustion, unburned hydrocarbons are oxidized with remaining oxygen in the cylinder. In addition to an adequate amount of oxygen, temperature and residence time have a strong influence on the oxidation rate. In only a few approaches [14, 73, 127, 181], chemical reaction kinetics are applied to describe the oxidation process. However, the validation for gasoline at engine conditions of high pressures and temperatures is very difficult. Furthermore, a detailed modeling of the combustion to define the concentrations of species including spatial and temporal resolution is required. This information cannot be provided by the presented simulation methodology in this work.

A more simplified approach to describe the oxidation of unburned hydrocarbons is chosen by Dent and Lakshminarayanan [16], who applied a threshold temperature of 1100K. Complete oxidation is assumed for an oxidation temperature above this limit value. At lower

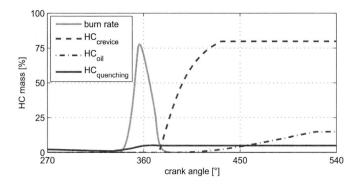

Figure 3.26.: HC emissions in the combustion chamber attributed to the individual sources without in-cylinder oxidation at cold engine operation of 1000min^{-1} and 2.5bar *bmep*

temperatures, no oxidation takes place and the calculated amount of HCs contributes to the engine-out emissions.

In the model presented here, an approach of one-step kinetics adapting an Arrhenius equation to gasoline introduced by Lavoie [93] is preferred as

$$\left(\frac{d\,[HC]}{dt}\right) = -Ae^{\left(-\frac{E_a}{RT}\right)}\,[HC]^a\,[O_2]^b \tag{3.77}$$

with the gas constant R, temperature T, and the concentration of remaining oxygen $[O_2]$. The calculation of the change in concentration of hydrocarbons $[HC]$ is adopted by other authors [66, 150, 156] and the parameters A, E_a, a, and b are fitted to iso-octane and other fuels. Different definitions of the constants and the investigated fuels are summarized in table 3.3.

Fuel	$A\left[\frac{cm^3}{mole}\right]$	$E_a\left[\frac{cal}{mole}\right]$	$a\,[-]$	$b\,[-]$	Source
gasoline	6.7×10^{15}	37230	1.0	1.0	Lavoie (1978) [93]
gasoline	7.7×10^{15}	37230	1.0	1.0	Schramm & Sorenson (1990) [150]
paraxylene	2.3×10^{14}	45040	0.882	0.657	Schramm & Sorenson (1990) [150]
propane	1.0×10^{23}	25000	1.0	1.0	Min & Cheng (1995) [114]
iso-octane	1.4×10^{10}	31000	1.0	0.25	Lavoie (1978) [93]
iso-octane	6.7×10^{15}	37230	1.0	1.0	Sodre (1999) [156]

Table 3.3.: Parameters of various fuels for the Arrhenius equation

Model calibration

It is clear from the presentation described above, that none of the phenomenological HC models discussed are predictive per se and certainly the model constants must be adjusted for the specific engine under consideration. The matching of the coefficients and parametrization is achieved by steady state operating points using measurements from an engine test bench. Due to the warmed-up operation conditions, only the models of piston crevice and oil layer are considered in the calculation of HC emissions. In the test engine, a Castrol *SAE 5W-30* oil similar to the one specified in [70] is used. The absorption/desorption of the fuel in the oil is based on the diffusion coefficient $D_{fuel,oil}$ that is defined by the equation of [179] and adapted to the parameters of the used oil. The Henry coefficient k_H is determined on the approach of iso-octane in *SAE 5W-20* by [157]. In the in-cylinder oxidation process, the Arrhenius equation with the parameters of gasoline according to [93] is calibrated with respect to the measurements by an additional multiplier comparable to [66, 150, 156].

Steady-state engine operation

The quality of the calibrated HC model is shown using a considerable set of steady-state engine operation points describing the whole map of a prototype, turbo-charged DISI engine with a fully-variable valvetrain (specified in table 4.1). The total amount of HC in the exhaust gas between cylinders and turbo-charger is analyzed by a flame ionization detector (FID). The relative proportion of the total HC emission by the formation sources crevice and oil layer and the fraction of the post-oxidation scaled to the measured amount are presented in figure 3.27. At this four exemplary points of steady-state engine operation, the mechanism of the oil layer ranges from 3 to 21% which can be confirmed by the determinations in the literature [12, 37, 41, 61, 66, 164, 182]. The in-cylinder post-oxidation rate of about 60% is also comparable to published studies [12, 126, 150, 164].

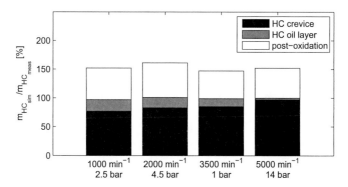

Figure 3.27.: Relative proportion of the HC sources and the post-oxidation to the total amount of formation compared to the measured value (100%) at defined engine speed and load (*bmep*)

In figure 3.28, the engine-out HC emissions relative to the measured value at 2000min^{-1} and 4.5bar *bmep* are compared between simulation and measurement at steady-state engine operation. For each speed, several points with increasing load (*bmep* spreads up to 1-18bar) are analyzed. At low values, the results of the model achieve a sufficient quality. It confirms the calibration strategy focused on this area of the engine map due to its importance for the implementation into the driving cycle simulation, since the engine mainly operates at low loads and speeds in the NEDC. Notable discrepancies can be seen at high loads and low engine speeds. Here, the internal exhaust gas recirculation is strongly reduced by scavenging. Moreover, considerable fractions of the post-oxidation process are delayed into the exhaust system. This results in the highest absolute values of engine-out HC emissions. Due to the assumption of a homogeneous stoichiometric mixture and the modal validity only in the inner-cylinder volume, these effects cannot be reproduced by the simulation model. Considering the extensive range of available data compared to other publications [16, 37, 66, 164, 182], the result achieves a similar quality over the whole engine map.

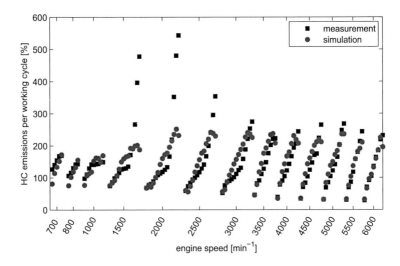

Figure 3.28.: Comparison of HC emissions between simulation and measurement at steady-state engine operation referred to 2000min^{-1} and 4.5bar *bmep*

3.7 TAILPIPE EMISSIONS

In section 2.4.4, an overview of the design and operation method of three-way catalysts is presented. According to the aforementioned requirements, the numerical simulation of the mechanisms in catalysts and the computation of conversion rates during quasi stationary light-off experiments are described in the following.

3.7.1 *Numerical simulation*

The determination of the conversation rates of three-way catalyst can be divided into the three modules channel flow, pore diffusion, and surface chemistry, respectively (Ullmann [167]). Each of them enables different modeling depths.

Channel flow

On the top-level, the exhaust gas flow through a single channel of the monolith is specified. Assuming a circular cross section in each channel, a symmetric profile for the laminar flow is considered. Thus, the two-dimensional Navier-Stokes equations (3.78 - 3.82) with the cylindrical coordinates z_r (radial) and z_a (axial) can be applied expressing the conservation of:

- mass:

$$\frac{\partial \left(\rho v_a\right)}{\partial z_a} + \frac{1}{r}\frac{\partial \left(r\rho v_r\right)}{\partial z_r} = 0 \tag{3.78}$$

- axial momentum:

$$\rho v_a\frac{\partial v_a}{\partial z_a} + \frac{1}{r}\rho v_r\frac{\partial \left(rv_a\right)}{\partial z_r} = -\frac{\partial p}{\partial z_a} + \frac{1}{\partial z_a}\left[\frac{4}{3}\mu\frac{\partial v_a}{\partial z_a} - \frac{2}{3}\mu\frac{\partial v_r}{\partial z_r}\right] + \frac{1}{r}\frac{\partial}{\partial z_r}\left[\mu r\left(\frac{\partial v_r}{\partial z_a} + \frac{\partial v_a}{\partial z_r}\right)\right] \tag{3.79}$$

- radial momentum conservation:

$$\rho v_a\frac{\partial v_r}{\partial z_a} + \frac{\partial}{r}\rho v_r\frac{\partial \left(rv_r\right)}{\partial z_r} = -\frac{\partial p}{\partial z_r} + \frac{\partial}{\partial z_a}\left[\mu\left(\frac{\partial v_r}{\partial z_a} + \frac{\partial v_a}{\partial z_r}\right)\right] + \frac{\partial}{\partial z_r}\left[-\frac{2}{3}\mu\frac{\partial v_a}{\partial z_a} + \frac{4}{3}\frac{\mu}{r}\frac{\partial \left(rv_r\right)}{\partial z_r}\right] \tag{3.80}$$

- species mass conservation:

$$\rho v_a\frac{\partial Y_k}{\partial z_a} + \frac{1}{r}\rho v_a\frac{\partial \left(rY_k\right)}{\partial z_r} = -\frac{\partial j_{k,a}}{\partial z_a} - \frac{1}{r}\frac{\partial \left(rj_{k,r}\right)}{\partial z_r} + M_k\dot{\omega}_k \tag{3.81}$$

- energy conservation:

$$\rho v_a\frac{\partial h}{\partial z_a} + \frac{1}{r}\rho v_r\frac{\partial \left(rh\right)}{\partial z_r} = v_a\frac{\partial p}{\partial z_a} + v_r\frac{\partial p}{\partial z_r} - \frac{\partial q_a}{\partial z_a} - \frac{1}{r}\frac{\partial \left(rq_r\right)}{\partial z_r} \tag{3.82}$$

According to a 2D boundary layer approach, the axial diffusion compared to the axial convective transport is negligible due to the boundary condition $ReSc\frac{L}{d} \gg 1$. In automotive exhaust converters typically, the Reynolds number Re is in a range of 10-300 (Ertl et al. [31]) and the Schmidt-number is about $Sc \approx 1$ (Bird et al. [6]). The surface reactions are linked to the gas phase by the diffusive flux in radial direction $j_{k,r}$. Thus, the second derivatives in the axial direction can be omitted. Further simplification is achieved by assuming a constant pressure in radial direction.

Since an infinitely fast diffusion in the radial direction is supposed, the radial gradients can be ignored resulting in a further reduction of the conservation equations. This effect implies a so called one-dimensional plug flow, where velocity, temperature, and species concentration are equal over the cross sectional area. Additionally, the momentum conservation equation in axial direction z_a can be eliminated defining a constant pressure in the channel due to no significant changes in exhaust converters (Ullmann [167]). The remaining conservation equations can be expressed as:

- mass conservation:

$$\frac{\partial \left(\rho v_a\right)}{\partial z_a} = \frac{4}{d}\sum_{k}^{N_G} F_{CatGeo}M_k\dot{s}_k \tag{3.83}$$

- species mass conservation:

$$\rho v_a \frac{\partial Y_k}{\partial z_a} = \frac{4}{d} F_{CatGeo} M_k \dot{s}_k - \frac{4}{d} Y_k \sum_k^{N_G} F_{CatGeo} M_k \dot{s}_k + M_k \dot{\omega}_k \qquad (3.84)$$

- energy conservation:

$$\rho v_a \frac{\partial h}{\partial z_a} = - \sum_k^{N_G} h_k M_k \dot{\omega}_k - \frac{4}{d} \sum_k^{N_G} F_{CatGeo} M_k \dot{s}_k h_k + \frac{4}{d} k_W \left(T_W - T \right) \qquad (3.85)$$

The right hand side of equation 3.83 becomes zero in steady state, when the system mass (molar mass M) is constant and the surface reaction terms ($\dot{s}$) are in balance. The diffusive mass flow on the surface is described by the product $M_k \dot{s}_k$. Since the reaction rates are specified for flat surfaces, the geometric factor F_{CatGeo} is required to increase the surface area due to the porous structure of the washcoat. In equation 3.84, the changes of each species (mass fraction Y_k) are described due to reactions between surface and gas phase (reaction rates $\dot{\omega}_k$). The temperature profile is determined by equation 3.85 with respect to heat release, consumption by chemical reactions, and heat exchange with the surroundings. The heat transfer between channel wall and gas phase is adapted by the coefficient k_W. The surface reaction rates $\dot{s}_k$ are effective values depending on the used pore and diffusion model (Ullmann [167]).

Pore diffusion

The next level is the modeling of the transport limitations in the washcoat. According to the literature, pore diffusion can be described by different approaches varying in complexity and computing time. The models range from detailed (3D pore network) [76, 136] over solving a reaction-diffusion equation (Hayes and Kolaczkowski [57]) to an analytic solution (effectiveness coefficients) (Chatterjee [11]). Due to computing time, a washcoat model using a 0D effectiveness factor is chosen. Since the radial concentration gradient in the washcoat is assumed as substantially greater than the axial one, the steady-state concentration profile of each species $c_k(z_r)$ in radial direction can be defined in a 1D reaction-diffusion equation as

$$\frac{\partial}{\partial z_r} \left(D_{eff,k} \frac{\partial c_k}{\partial z_r} \right) + \gamma \dot{s}_k = 0 \qquad (3.86)$$

with the effective diffusion coefficient $D_{eff,k}$ and the geometric parameter γ as function of catalytic surface to washcoat volume:

$$\gamma = \frac{A_{catalytic}}{V_{washcoat}} = \frac{F_{CatGeo}}{L_{pore}}. \qquad (3.87)$$

The length of the pore is defined by L_{pore} and is assumed equal to the washcoat thickness in this work. An equal concentration of each gas-phase species is defined at the pore entrance and at the gas-washcoat interface. Thus, following connection to the plug flow approach is implemented:

$$M_k F_{CatGeo} \dot{s}_k = M_k D_{eff,k} \frac{\partial c_k}{\partial z_r} \bigg|_{z_r=0}. \qquad (3.88)$$

Following assumptions are required to solve equation 3.86 analytically for a single species (Hayes and Kolaczkowski [57]):

- The diffusion coefficient D_{eff} is constant.

- The species is consumed and the reaction rate $\dot{s}$ is proportional to the concentration c.

- The concentration gradient at the washcoat to the wall interface is negligible compared to the thickness of the washcoat layer.

According to the defined conditions, it results the analytical solution of the reaction diffusion equation

$$c\left(z_r\right) = \frac{e^{-\varsigma}}{e^{\varsigma} + e^{-\varsigma}} c_k \cdot e^{\frac{\varsigma z_r}{L_{pore}}} - \frac{e^{\varsigma}}{e^{\varsigma} + e^{-\varsigma}} c_k \cdot e^{-\frac{\varsigma z_r}{L_{pore}}} \qquad (3.89)$$

with the Thiele modulus ς as ratio of reaction rate to diffusion rate (Ullmann [167]):

$$\varsigma = L_{pore} \sqrt{\frac{\gamma k}{D_{eff,k}}}. \qquad (3.90)$$

The rate coefficient k is determined by applying the Arrhenius approach:

$$k = AT exp\left(-\frac{E_a}{RT}\right). \qquad (3.91)$$

Surface chemistry

On the last level, the chemical reaction mechanisms in the pores of the washcoat are established. For three-way catalysts in automotive applications, processes in the gas phase can be neglected and only reactions on the catalytic active surface area have to be considered (Koop [86]). The characteristic processes on the catalytic surfaces can be classified into three steps (Mladenov [115]):

Adsorption: Based on the high temperatures in the catalyst, the chemisorption leads to high adsorption enthalpy due to the chemical bond between adsorbate and absorbent. This can result in dissociation of the bonded molecules, called dissociative adsorption. In figure 3.29(a), both adsorption phenomena are illustrated. Small activation energy is characteristic for the adsorption process.

Reaction: Two significant mechanisms can describe the reaction kinetics of adsorbed species:

- The *Langmuir-Hinshelwood* mechanism assumes a directly reacting of the adsorbates to each other, as shown in figure 3.30(a).

- The *Eley-Rideal* mechanism describes the chemical reaction of an adsorbate to species in the gas phase and the desorption of the new formed species from the surface afterwards (presented in figure 3.30(b)).

Desorption: The phenomenon desorption can be depicted analogously to the adsorption process. Thus, simple or associative desorption can occur as illustrated in figure 3.29(b).

Simulation of three-way catalytic converters are investigated in multiple studies introducing several mechanisms with different levels of detail [38]. Depending on the complexity of the mechanisms they can be classified into three main categories:

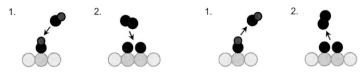

(a) simple adsorption (1.) and dissociative adsorption (2.)

(b) simple desorption (1.) and associative desorption (2.)

Figure 3.29.: Adsorption and desorption processes [115]

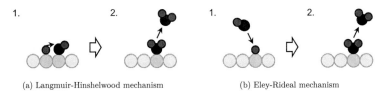

(a) Langmuir-Hinshelwood mechanism

(b) Eley-Rideal mechanism

Figure 3.30.: Chemical reaction mechanisms of adsorbed species [115]

Detailed reaction mechanisms: The chemical reaction kinetics is expressed on a molecular level by elementary-like reaction steps including the separated phenomena of adsorption, reaction, and desorption. This requires the definition of reaction equations for species in the gas-phase and on the surface. Additionally, pore diffusion has to be considered.

Global reaction mechanisms: In order to reduce computing time for transient simulations, the detailed chemical reaction kinetics can be simplified by defining global reaction mechanisms. Only a few fundamental equations have to describe the chemical reactions of the main pollutants. Often, these mechanisms were expanded by additional equations to consider transient phenomena of oxygen storage.

Simplified models based on elementary step kinetics: These models use reaction mechanisms based on reduced elementary step kinetics and simplified calculations of the reaction rates.

A comparison of various modeling approaches presented in the literature is listed in table A.1. Due to its well-balanced proportion of physical background and usability, the global reaction mechanism of Holder et al. [64] (described in the appendix A.3) is implemented into the simulation model for all investigations presented in this work.

In summary, figure 3.31 shows an overview of the three modules described above including the required input parameters and calculated results. With respect to computing time and available data of measurements, the approaches of simplified models were applied (represented underlined in figure 3.31) in all modules. The reactions on the surface of the washcoat are expressed by a global reaction mechanism, while mixed diffusion (Knudsen and molecular diffusion) is assumed inside the pores realized by an effectiveness factor. The flow through the channel is simulated by an approach of isothermal 1D plug flow.

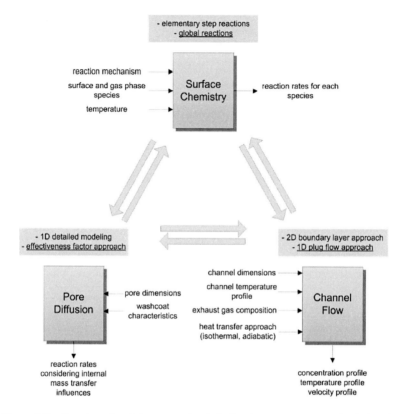

Figure 3.31.: Overview of modular approaches including input parameters and calculated results [38]

3.7.2 Simulation of the light-off behavior of three-way catalysts

If the three-way catalyst is heated up to its operation temperature and the combustion of the engine is controlled at stoichiometric mixture, the conversion rates achieve nearly 100% for the three key emission components CO, NO_x, and HC. The important fact in the exhaust aftertreatment to reduce tailpipe emissions is the improvement of the light-off behavior of catalytic converters. In simulation models furthermore, the temperature-dependency of the washcoat chemistry has to be reproduced at first to predict transient behavior in driving cycles.

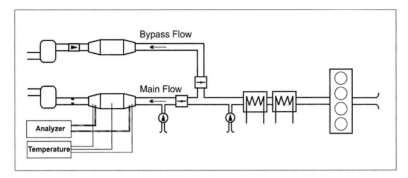

Figure 3.32.: Experimental set-up to measure light-off temperatures of TWC's [63]

The light-off temperature is defined by the conversion rate of 50% and can be determined by specialized measurements. The experiment set-up (illustrated schematically in figure 3.32) consists of an engine operating in steady-state and producing a quasi-stationary exhaust gas flow of approximately stoichiometric and constant gas composition. The exhaust gas temperature is cooled by heat exchangers initially to ambient conditions and then adjusted to a very small temperature gradient lower than 0.5K/s. Due to this conditioning of the exhaust gas flow, a quasi-stationary state and a homogenous temperature distribution inside the catalyst can be assumed. The total conversion of the relevant species (CO, NO_x, and HC) is calculated by measuring the gas composition at the entrance and exit of the catalytic converter:

$$r_{conv\_k} = \left(1 - \frac{c_{k,out}}{c_{k,in}}\right) \cdot 100 \qquad (3.92)$$

The simulated light-off behavior is validated by measurements from a commercially available three-way catalyst build for a 4-cylinder SI engine. It is a two-brick configuration with the same palladium/rhodium loading in both segments. The main parameters and specifications are presented in table 3.4. Corresponding measurements are also conducted with this type of catalyst to analyze the influence of both, aging and precious metal loading.

The simulation of the light-off characteristics can be splitted into three process steps (figure 3.33). In the preprocessing, the data from the measurements are prepared for the actual calculation by setting the sampling conditions. Combined with the parameters of the specific catalyst, the objects of plug, pore, and surface are created. After the calculation of the complete temperature profile, the deviation of the simulation results to the measurements is determined in the post-processing.

Global reaction mechanisms, introduced above, describe only one certain washcoat depending on its material, support, structure, and method of manufacturing (Holder [63]). For different washcoat properties (e.g. various precious metal loading), the kinetic parameters of the chemical reactions have to be adapted to the catalysts used for validation. Therefore, an automated optimization process including the investigation of different optimization algorithms is implemented (Frommater [38]). The results of the conversion rates of the main pollutants CO, NO_x, and HC versus temperature are shown in figure 3.34. Compared to the measured characteristics, the light-off temperature is matched very well. Only minor deviations occur at high conversions rates of CO, for example.

parameter	brick 1	brick 2
geometric data		
monolith length [m]	0.084	0.090
monolith diameter [m]	0.11	0.11
cell density $[\text{cells/in}^2]$	600	400
wall thickness [mil]	3.5	4.3
total loading $[\text{g/ft}^3]$	120	50
cell diameter $[\mu m]$	90	110
washcoat data		
porosity	0.3	0.3
pore diameter [nm]	12	12
pore length $[\mu m]$	97	104
turtuosity	3	3
F_{CatGeo}	100	100
flow data		
pressure [hPa]	100	100
mass flow rate [kg/h]	110	1100

Table 3.4.: Parameter settings of the catalyst to investigate light-off behavior [38]

3.7.3 *Influence of aging*

Three-way catalysts can degrade with respect to their conversion behavior due to aging effects. This gradually process can be categorized by its three sources: chemical, thermal, and mechanical aging [4, 92]. Especially, the chemical and thermal effects are crucial for the catalyst deactivation under real driving conditions [84, 102].

Thermal aging causes sintering of the precious metals and/or the washcoat resulting in a reduced catalytically active surface area. According to Mladenov [115], these processes are illustrated basically in figure 3.35. At high temperatures, the precious metals become movable and volatile inside the washcoat forming particles of larger size, called metal sintering (Holder [63]). Since the washcoat component gamma alumina (Al_2O_3) undergoes phase changes with lower surface areas above a critical temperature, sintering of the washcoat occurs. Both processes of thermal sintering arise at temperatures above 500°C and can be intensified by water vapor of the exhaust gas.

Chemical aging describes the blocking of active sites and the reducing of binding strength in the catalytic converter by poisoning elements from fuel and engine oil like sulfur, phosphorus, and magnesium (Angelidis and Sklavounos [2]). In catalysts used for gasoline engines, the main chemical aging effect is caused by oil additives. Fuel components, such as lead and manganese, are also strong poisons, but can be neglected in countries with highly restrictive emission regulations due to the high quality of fuels. Physical damage of the catalyst can occur by outside influences, e.g. hitting the ground while driving or corrosion of the metal housing. Usually, mechanical aging can be excluded. More detailed information about the different aging mechanisms is explained in (Lanzerath [91]) and (Jobson et al. [72]). Summarizing, the primary process for catalyst deactivation or malfunction is nowadays thermal aging due to the reduction of oil consumption in modern engines and the improved fuel quality (Ignatov et al. [67]).

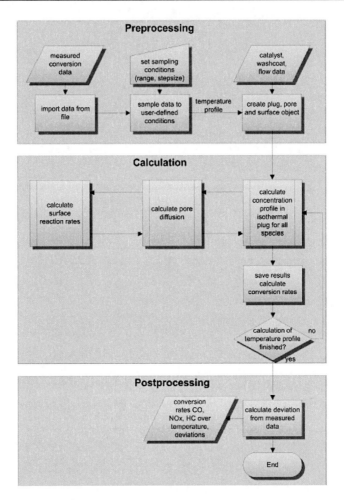

Figure 3.33.: Process of calculating light-off characteristics of TWC's [38]

In figure 3.36, the deactivation impact by aging is demonstrated at the light-off character-istics of CO, NO_x and HC. As shown, the measured light-off temperature of all pollutants is shifted to almost 100K higher values for the aged catalyst. The aging mechanisms can be simplified described as an apparent reduction of available reactive surface area, while the in-trinsic heterogeneous reactions persist unchanged (Holder [63]). Since the inhibition term and the activation energy of the chemical reactions remain the same, the resulting reaction rate of an aged catalyst can be scaled by a specific surface area F_{CatGeo} (known from Deutschmann

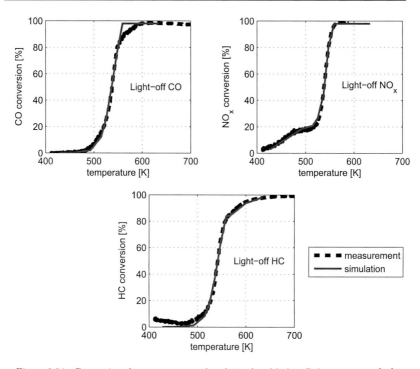

Figure 3.34.: Comparison between measured and simulated light-off characteristics [38]

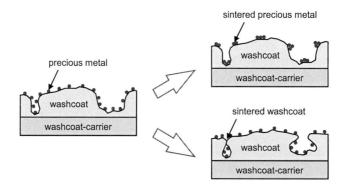

Figure 3.35.: Processes of thermal aging causing sintering of the precious metals and the washcoat [115]

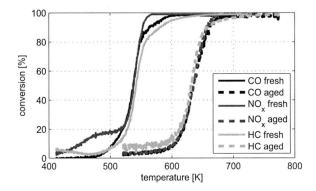

Figure 3.36.: Deactivation impact by aging to the conversion of CO, NO_x and HC [38]

et al. [20]). The scaling of the active surface area also affects the effectiveness factor in the pore diffusion (see equation 3.94).

In this validation process, the measurement of the aged catalyst is set as reference and the simulation model is matched to it by optimizing the parameters (as described above). A value of one is defined for the F_{CatGeo}^{aged}, since no information about the catalytically active surface area is provided. This approach assumes a catalytic surface area equal to the geometric one. Another simplification is the definition of a constant change in the active surface area to avoid a speculated aging profile (Holder [63]).

$$k^{aged} = k^{fresh} \cdot \frac{F_{CatGeo}^{fresh}}{F_{CatGeo}^{aged}} \tag{3.93}$$

$$\gamma^{aged} = \gamma^{fresh} \cdot \frac{F_{CatGeo}^{fresh}}{F_{CatGeo}^{aged}} \tag{3.94}$$

The light-off behavior of the fresh catalyst is simulated by using the same parameter set of the aged one and optimizing only the ratio of $F_{CatGeo}^{fresh} / F_{CatGeo}^{aged}$. The simulation results compared to the respective measurements are presented in figure 3.37. The specific surface area of the fresh catalyst F_{CatGeo}^{fresh} is now 10.47. It means that the aged catalyst lost his conversion performance by more than 10 times of the primarily catalytically active surface. While the simulated light-off characteristics match very well for the aged catalyst, some deviations at low and high temperatures occur for the fresh one due to the simplifying assumptions.

Summarized, the validation shows a good reproduction of the light-off temperature at different aging stages by adjusting only one parameter of the model. This makes the handling of the simulation easy, since all reaction rates are scaled similarly. However, further investigations about effects on the conversion behavior by various aging conditions are required to determine a correlation of F_{CatGeo}.

3.7.4 Influence of precious metal loading

In the design of three-way catalysts, the precious metal loading plays a major role affecting significantly the conversion behavior and costs of the component. In the presented investi-

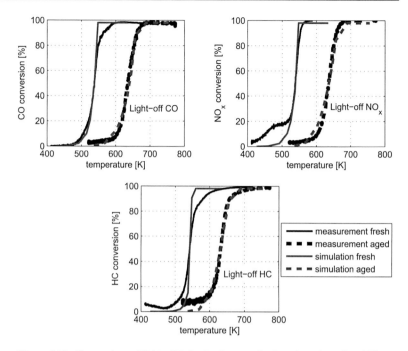

Figure 3.37.: Comparison of light-off behavior between fresh and aged catalyst [38]

gation, the absolute amount of precious metals is varied, whereas the Pd/Rh ratio remains constant.

The consequences of changing the precious metal loading are comparable to the processes in the washcoat during aging (Deutschmann et al. [20]). Kang et al. [74] introduce an activity function for the active surface area to describe the influence of noble metal loading on the reaction rates. The varying loading (*load*) is incorporated by a global parameter a applied to the Arrhenius equations:

$$k^{load} = a \cdot k^{140g/ft^3} \tag{3.95}$$

The simulation model is calibrated by the same parameter set of the aged catalyst from the previous investigation (section 3.7.3), since the corresponding measurements of this analysis are also conducted with aged converters. The reference loading of precious metals is $140g/ft^3$ $\left(k^{140g/ft^3}\right)$ with the factor $a = 1$. The total load is varied in both directions, reduced to $100g/ft^3$ and raised to $170g/ft^3$, while the Pd/Rh ratio remains constant. A comparison of the loading influence is presented in figure 3.38 including the validation of the simulation results. The adaption of the Arrhenius rate equations shows a good agreement of the conversion rates with the measured characteristics. The deviations at lower temperatures occur already at the reference loading and cannot be compensated by a global parameter.

The values of the derived factor a corresponding to the precious metal loading are shown in figure 3.39. Even only three different total loadings are compared, a trend of a non-linear

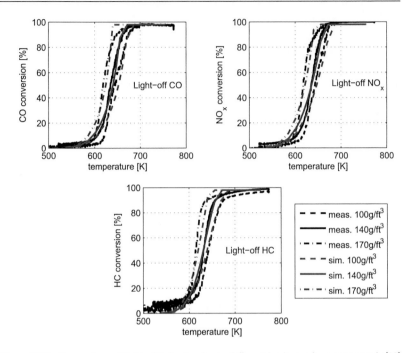

Figure 3.38.: Comparison of light-off behavior due to different loadings of precious metals [38]

parabolic curve rising steeper with higher loadings can be identified. However, this result is only valid for the considered aging conditions. Additional investigations are required to validate this function for different catalysts.

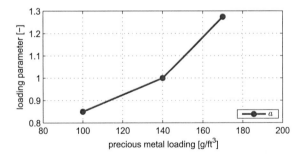

Figure 3.39.: Values of the factor a corresponding to the precious metal loading of an aged catalyst [38]

4

TRANSIENT SIMULATION

After describing the modeling of the coupled simulation methodology, its application under transient operation and the validation of its results are presented in this chapter. Several measurements are conducted to investigate the quality of each sub-model and the coupled overall simulation. A compilation of the variations and the specification of engine and vehicle the matching is based on are considered in the first section 4.1. A validation process has to be established to reduce the complexity of the closed-loop system by a structured and standardized proceeding (4.2). The results of the matching are shown and discussed in the proof of concept (4.3).

4.1 MEASUREMENTS

In order to guarantee a solid validation of the introduced simulation methodology, various influences to prove specific parts of the coupled overall system are investigated both, experimentally and numerically. The matching is carried out using measurements of entire driving cycles. Figure 4.1 shows an overview of the variations and their correlations. For the demonstrated validation shown in this chapter, usually, the speed profile of the NEDC or parts of it is used exemplarily. However, a random driving cycle called RCA to examine Real Driving Emissions (RDE) is used for further investigations as well. It exhibits more dynamic characteristics with regard to matching the engine model at higher loads and testing the behavior of the driving controller, respectively.

The reference measurement (ref.) is represented by a calibration based on a standardized parametrization of the ECU, still in development status, but already aiming for optimizing fuel consumption and emissions. The entire driving cycle is performed with the engine at operating temperature, thus turning off effects of warm up. However, driving cycles regulated by law have to be cold started. This requirement is observed in additional tests of both vehicle speed profiles with a conditioned engine starting the driving cycle with oil and coolant temperature at ambient state (cold). Hence, it also considers the influence of catalyst heating. Further variations to prove the functionality of particular sub-models consist of driving cycles, all started with a warmed-up engine, with respect to individual modifications:

The gas exchange of the engine model is analyzed by modifying the calibration of the ECU concerning the actuation of valves and throttle. Compared to the reference, less valve overlap timing is applied resulting in a significant increase of fuel consumption and engine-out emissions of NOx (var.1). In a further measurement (var.2), the variable valve lift is deactivated and the intake valves open with maximum lift at all times. The engine load is now controlled by the throttle. Specific variations are conducted to validate the results of the engine-out CO and HC emissions. Therefore, the air-fuel ratio is changed from stoichiometric to fuel-rich (var.3) and lean (var.4) mixture preparations.

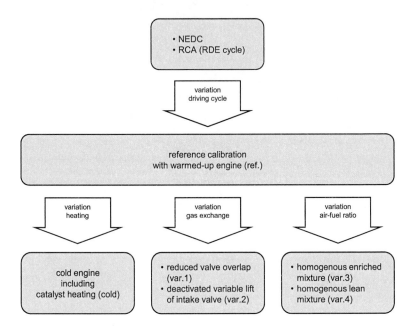

Figure 4.1.: Overview of the diversified measurements

The presented measurements are performed at a highly-dynamic engine test bench providing a good reproducibility (Froschhammer et al. [39]). Another advantage is the well controlled simulation of the vehicle behavior analogical to the coupled overall model. Thus, discrepancies/tolerances of the roller chassis dynamometer can be prevented. At the test bench, a prototype of a four cylinder DISI engine with state-of-the-art technology features was installed. It is equipped with direct injection and a full variable camshaft system to control the lift of the intake valve and cam timing at both, inlet and outlet. High boost pressures are achieved by a twin-scroll exhaust-gas turbocharger. The virtual drive train is transferred from a vehicle of the premium small car segment with manual transmission (MT) and front-wheel drive (FWD). The main specifications of engine and vehicle are summarized in table 4.1.

4.2 VALIDATION PROCESS

The coupled overall simulation model with the different sub-models is a complex closed-loop system. In order to reach a successful validation, it is necessary to implement a defined process (illustrated in figure 4.2). First of all, the single sub-models have to be analyzed separately, before they can be integrated into the coupling. This is important, because each sub-model has its own characteristics and needs specific criteria for plausibility checks. The individual parts of the validation process are specified in the following and consequently, the procedural method in the upcoming section 4.3 is described.

Engine	in-line 4 cylinder
cubic capacity	1998cm^3
bore	82, 0mm
stroke	94, 6mm
compression ratio	10, 2
maximum power	170kW at $5200 - 6000$min^{-1}
maximum torque	320Nm at $1250 - 4800$min^{-1}
idle speed	700min^{-1}
Vehicle	MT FWD
acceleration $0 - 100$km/h	6.3s
maximum speed	246km/h
unloaded weight	1280kg

Table 4.1.: Specification of engine and vehicle according to BMW Group [8]

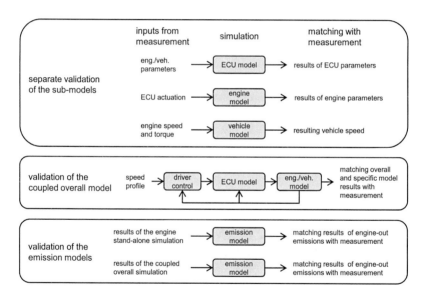

Figure 4.2.: Steps of the validation process

The aim for the model of the ECU is to achieve an envisaged balance between functional quality and modeling costs. In the validation process, all required signals coming from the engine or vehicle are set by a measurement from the engine test bench. Subsequently, the calculated actuators in the model can be compared with the corresponding measurement parameters. This procedure also allows the validation of separate modules of the ECU to investigate particular functions of interest.

The sub-model of the combustion engine is verified in two steps. First, a validation of engine performance is conducted in steady state operating points using measurements from an engine test bench (already shown in chapter 3.3). In the second step, the validation procedure of the transient combustion engine simulation follows the same principle as for the ECU process mentioned above. But here, the actuators coming from the real ECU are taken from a measurement and those virtual sensor parameters which are sent from the engine to the ECU model are matched against the measured reference values. A matter of particular interest is the quality of gas exchange in the engine model, because of the strong correlation between fuel consumption and air flow.

The vehicle powertrain sub-model is validated using the measured vehicle speed at a defined incoming engine speed and torque signal from the ECU. This approach proves not only the kinematic transformation of the powertrain, but also its kinetic mechanisms. The vehicle powertrain is investigated as one global system and the particular components (e.g. gearbox, differential etc.) are not studied individually. However, a more detailed verification is not required for the purposes of this simulation methodology.

In the final step of the validation process, the interaction of the coupled sub-models together with the virtual driver is inspected. The overall model represents a closed-loop simulation, where incoming parameters of each sub-model come from another one. This means that deviations of one sub-model can lead to divergences in the next sub-model(s) and therefore in any further calculation of the overall model. The parameters of interest are the same as used in the previous validation process.

Since the determination of the engine-out emissions is implemented in the post-processing of the engine simulation, the validation of the results is also a subsequent process step. Depending on the engine simulation, individually or coupled, the emission models can be verified in detail initially and are then matched on the basis of the whole system. The matching of the engine-out NO_x and CO emissions is conducted by analyzing the relative difference between the cumulated profiles of the reference calibration and the particular variation (as illustrated in figure 4.3).

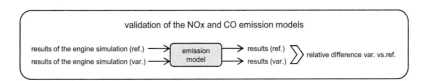

Figure 4.3.: Validation process of the NOx and CO emission models

In most cases, the matching of the results is presented as comparison between the normalized profiles of the measurement $meas_{rel}(t)$ and simulation $sim_{rel}(t)$ during the driving cycle. The relative values are calculated as

$$meas_{rel}(t) = \frac{meas(t)}{max(meas)} \cdot 100\% \tag{4.1}$$

$$sim_{rel}(t) = \frac{sim(t)}{max(meas)} \cdot 100\% \tag{4.2}$$

where the maximum of the entire measured profile $max\,(meas)$ is the reference. Parameters with continuously negative values (e.g. $tq\_friction$) are defined as

$$meas_{rel}\,(t) = \frac{meas\,(t)}{-min\,(meas)} \cdot 100\% \qquad (4.3)$$

$$sim_{rel}\,(t) = \frac{sim\,(t)}{-min\,(meas)} \cdot 100\% \qquad (4.4)$$

with the minimum of the entire measured profile $min\,(meas)$. Parameters with a profile including positive and negative values are referred to the maximum measured peak.

4.3 PROOF OF CONCEPT

According to the introduced validation process, an exemplary subset of transient results is presented and discussed in the following. Each sub-model of ECU, combustion engine, and vehicle are matched, as well as the coupled simulation environment. Complications and a path of error summation are identified for the closed-loop simulation. Furthermore, the emission sub-models of NO_x, CO and HC are validated by measurements and a short summary is given.

4.3.1 ECU model

According to section 4.2, the validation starts with the separated ECU model. The aim of this sub-model is to obtain the desired quality in the calculated actuator quantities for the output to the combustion engine. Therefore, the complete path from the driver request to the actuator modules has to achieve a good reproduction of the real ECU. Particular modules or the entire ECU model can be tested by stimulating the required inputs with measured signals.

Figure 4.4 illustrates several results of important interface parameters in the torque module according to main path variables described in chapter 2.1.1. From top to bottom, the first comparison shows the desired torque calculated on the basis of the acceleration pedal position from the driver. This is charged with losses of the engine and auxiliary torque demands by environment conditions and vehicle functions, here exemplified by the friction torque and an additional request from the catalyst heating module, respectively. Latter one is a fundamental part for the prediction of fuel consumption in driving cycles with cold started engines due to the changed operation strategy. It can be stated that the simulation results of the sub-modules agree very well with the measurement. This can be approved by the desired indicated torque following from the torque charging. While the spark timing is based on the desired torque, the parameters from the air path correspond to the air flow. Therefore, the torque demand is converted in the ECU. Its comparison between simulation and the measured calculation is presented in the bottom plot. Summarized, the torque module achieves the requested quality and provides the required information to the control modules of spark and air flow.

Beside the good results of the torque demand from the torque module, the ECU model has to achieve a good quality in the matching of the actuator parameters transferred to the combustion engine. The validation of the output results are represented in figure 4.5. At the top, the lift of the intake valve is reproduced by the simulation model with high analogy. Some deviations can be detected at the actuation of the valve timing. These parameters are influenced by a function based on the detection of irregular engine running (e.g. due

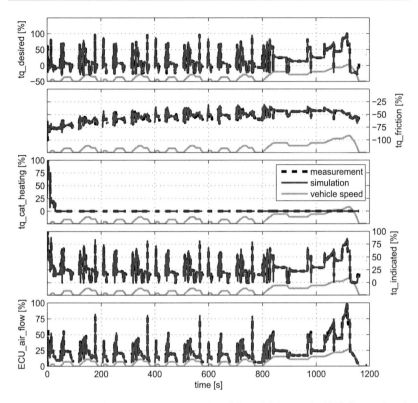

Figure 4.4.: Results of the torque module in the ECU model during the NEDC started with cold engine

to occasional misfire events). The ECU monitors this by checking the rotation speed of the crankshaft, but the required sensor signal is not available in the present measurements. Nevertheless, the quality of the actuation of the intake valve meets the demands. The cam timing of the exhaust valve is not that effected resulting in a good matching. The spark timing has to fulfill high standards in response time and accuracy of the values due to its strong impact on the combustion process and engine torque, consequently. However, the output of the spark module shows a high conformity to the measurement. Altogether, the achieved quality of the actuation parameters approves the application of the ECU model to control air flow and spark timing.

In figure 4.6, the actuator parameters of the ECU model for the variation with the changed valve overlap timing (var.1 from section 4.1) are validated. Smooth adaptions are implemented to the valve lift by the engine control. However, the difference in the actuation strategy compared to the standard calibration can be detected clearly at the profiles of the valve timing of both, intake and exhaust. In the reference NEDC, fast switches between high and low overlaps are performed. Here, the valve timing is nearly constant and the adjustment

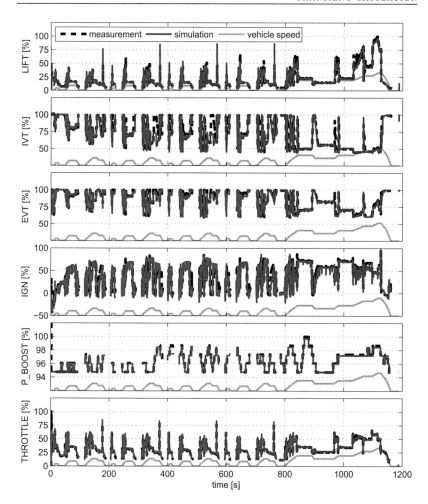

Figure 4.5.: Results of the actuator parameters calculated in the ECU model during the NEDC started with cold engine

can be achieved in a simpler way. Due to the small changes in the valve actuation, effects of irregular engine running can be neglected in this case. Since the gas exchange is modified, the spark timing is controlled according to the changed mixture preparation. No variations can be observed in the control of the boost pressure, thus it is skipped in the shown matching. Compared to the measured characteristics, the simulation results achieve a high congruency.

The validation approves the modeling depth of the ECU functions. Even the individual modules are simplified, the calculations achieve the required quality. In summary, the ECU

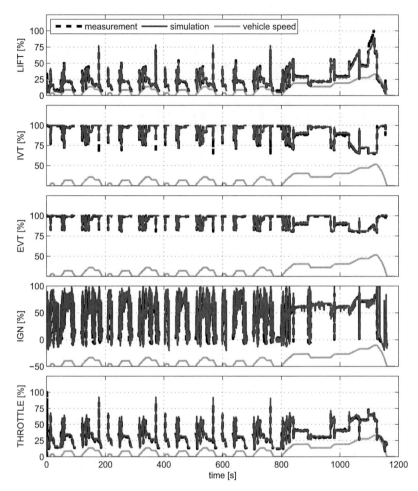

Figure 4.6.: Results of the actuator parameters calculated in the ECU model for the changed valve timing strategy (var.1) during the NEDC

model conforms all demands from 2.1.1 and can be integrated into the coupled overall simulation.

4.3.2 Engine model

As described in section 4.2, the transient behavior of the virtual combustion engine has to be proven (see chapter 3.3 for steady-state validation). Therefore, engine speed as well as the actuating parameters from the ECU are provided as defined input by measurements for the simulation results in the following.

In figure 4.7 several important parameters of the gas exchange in the engine model along the entire NEDC are shown. Since the engine speed is a predetermined input from the measurement in this simulation, it consequentially matches accurately (as seen on the top). A good accordance of the gas exchange behavior with the results of the engine test bench is recognizable by the characteristics of pressure in the intake manifold (man. pres.) and air flow. Only during the extra-urban part, some discrepancies can be observed in the manifold pressure. These are required to achieve the high air flow rates at the end of the speed profile. Corresponding to the direct influence of the air mass, the cumulated CO_2 emissions show a good correlation. This is accomplished not only for the absolute value at the end of the driving cycle, but also for the complete curve progression. It has to be mentioned that deviations occurring during the driving cycle have an impact on the remaining profile due to the accumulation of the values.

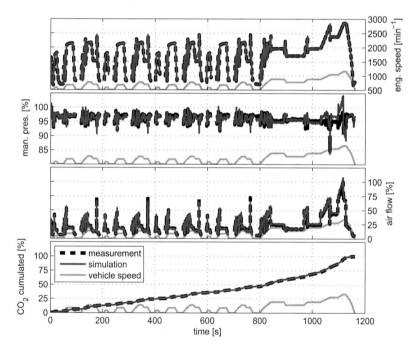

Figure 4.7.: Results of several parameters in the engine model during the NEDC started with a warmed-up engine

Figure 4.8 exhibits a more detailed look at the process of gas exchange and how it is affected by variations of the calibration (var.1 and var.2 explained in section 4.1). Hereby, an important advantage of this simulation methodology with a detailed engine model can be observed (Dorsch et al. [25]). Since every working cycle of the combustion engine is calculated individually, a user-defined zoom into the driving cycle is permitted. Illustrations of the pressure profile in intake and exhaust tract and the air-mass flow through intake and exhaust valve are arranged in pairs for each variation. On the top, the reference (ref.) shows the air conditions at around 1079.5s of the NEDC with the standardized parameterization. The differences in the strategy of the calibration can be identified by the graph of the valve lifts and intake pressure, respectively. At the first variation (var.1), the valve overlap is reduced significantly. While the intake manifold pressure is nearly unchanged, slight deviations can be detected in the exhaust manifold. More influence of the valve timing can be seen at the air flow. The peak of the air mass through the intake valve is flattened and the reverse flow from exhaust into the cylinder is prevented almost completely. This has a considerable impact to the residual gas content. Compared to the reference calibration, it is decreased from 13.8% to 8.6%. In var.2, the variable valve lift is disabled and the intake valves open with maximum lift at each working cycle. Due to the changed opening, the valve timing is adapted to less overlap by the ECU. The engine load is now controlled by the throttle resulting in a clearly lower intake pressure in the presented period. The pressure in the exhaust manifold is comparable to var.1. A different characteristic of the mass flow rates through the valves, especially on the intake side, can be observed and a residual gas content of 8.1% similar to var.1 can be detected. In summary, this example confirms that effects of valve control to the gas exchange behavior by different calibration strategies can be depicted and investigated by this simulation methodology.

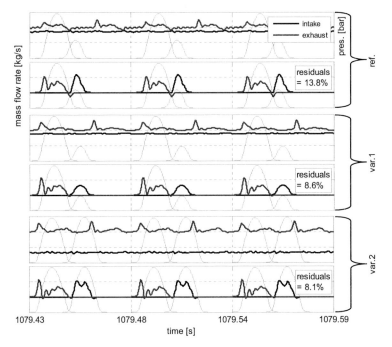

Figure 4.8.: Comparison of gas exchange due to variations in calibration strategies

In order to make predictions of various concepts with respect to valve actuation, the engine simulation has to produce qualitative results over a wide range of calibration under non-steady operating conditions. For the purpose of this work, several investigations to improve the transient gas exchange performance are conducted. Among other parameters, the air flow in the engine model can be influenced significantly by adjusting the discharge characteristics of the throttle and valves (introduced in chapter 3.3.1) [48, 83, 140]. In figure 4.9, two different

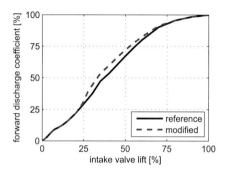

Figure 4.9.: Example of different discharge coefficients in forward direction at the intake valve referred to the maximum value

profiles of the discharge coefficients at the intake valve in forward direction are shown to exemplify the calibration of the air mass flow. The graph can be divided into three parts. In the lower section within 25% of the maximum lift, engine operation at idle speed and during constant vehicle speed periods in the NEDC can be matched. The range beyond 75% valve lift belongs to maximum opening actuation, e.g. during full load operation. The lifts in between effect the mass flow rates during accelerations and are important for the determination of fuel consumption in driving cycles. For example, an averaged increase of not even 10% of the discharge coefficients in this middle lift range (modified profile) results in 1-2% higher CO_2 emissions dependent on the strategy of valve actuation during the NEDC. However, the flow characteristics of the valves cannot be modified at will. Flow directions and rates have to show a realistic behavior and can be controlled by detailed analyses like in figure 4.8.

In this work, one parameter setting including a compromise between the variations of engine calibration introduced above is optimized for both driving cycles. Based on the investigations, most effective is the adaption of the intake forward and exhaust reverse air flow. The discharge characteristics of the throttle and the upper range of valve lift are mainly verified by the measurement of var.3. Challenging is the matching to different strategies of valve actuation like the differing valve overlap between the engine calibrations ref. and var.1. Especially during the time period, when intake and exhaust valves are open at the same time, the determination of cylinder flows seems to be critical. For this purpose, the adjustment of the reverse coefficients at the exhaust valve has an useful impact. However, an opposing trend of the air mass can be detected for both calibration variations during the matching. It is supposed that the determination of the cylinder flow is limited due to the modeling approach in the 1D environment. Hence, further investigations about the valve flow characteristics to develop new calculation methods like [22, 104] are required in the future.

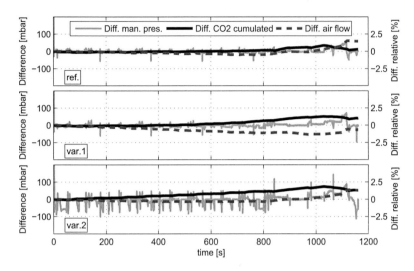

Figure 4.10.: Validation of gas exchange and fuel consumption due to variations in calibration strategies during the NEDC

The model adjustments are verified by comparison of the simulation results with the corresponding measurement for each of the three different engine calibrations. The gas exchange and fuel consumption is validated in figure 4.10 by plotting the cumulated divergence of air flow and CO_2 and absolute deviations of the manifold pressure for the entire NEDC. Overall, the matching of the simulation shows a good accordance to the measured behavior and confirms the adaptions of the engine model. In all three variations, the largest differences of the air flow are located within the limits of $+/-2.5\%$ referred to the end value of the measurement. The relative discrepancies of the cumulated CO_2 profiles do not exceed this range of tolerance, as well. In the case of full variable valve timing (ref. and var.1), the manifold pressure can be reproduced with small deviations by the simulation model. In var.2, the lift of the intake valve is fixed to maximum opening and the engine is throttled only by the throttle resulting in a considerably more dynamic profile of the manifold pressure, especially at accelerations. This explains the higher oscillations of the absolute difference. It is noticeable that the largest divergences can be detected at the end of the NEDC for all three calibration strategies. As mentioned above at the validation of several parameters in the engine model, this is accepted to achieve a better matching of the air flow. At this period of the driving cycle, it can also be observed a partial opposing trend between the profiles of air flow and CO_2. This indicates an incorrect reproduction of the gas exchange in these engine operation points. Nevertheless, the considerably slight fluctuations in the differences enable the application of the engine model within the coupled simulation environment to investigate transient conditions.

Further validation regarding transient performance of the engine model is studied in the more dynamic driving cycle RCA. In Figure 4.11, the same simulated parameters as shown for the NEDC are matched to measurements of an entire RCA started with a warmed-up engine. Here, the vehicle speed profile is more aggressive resulting in a larger oscillation of the engine speed. Due to the operation with full variable valve actuation, the manifold pressure is almost on a constant level that can be reproduced by the simulation. Although the air flow has a dynamic profile, the deviations of the engine model can be neglected. It seems that the gas exchange at higher engine speed and load is predicted accurately. This can be confirmed by the high correlation ratio of the cumulated CO_2 emissions.

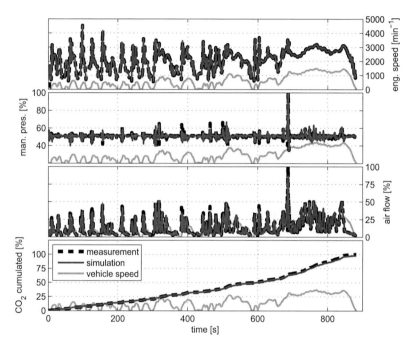

Figure 4.11.: Results of several parameters in the engine model during the RCA started with a warmed-up engine

In the appendix A.1.3, further results of both driving cycles, NEDC and RCA started with a cold engine are presented. Due to the difficult conditions and activated catalyst heating, the simulation is more complex here. However, the engine model achieves accurate predictions of the gas exchange and fuel consumption. The matching confirms the compromise of this work between a detailed level of modeling and the approach of applying friction maps for different temperatures.

The validation verifies the desired quality of the engine model at transient operation conditions. A good accordance to measurements is accomplished for different calibration strategies and engine conditions during various driving cycles. The matching of the air flow shows some discrepancies in the gas exchange due to the large range of valve timing actuation. Here, further investigations are required to improve the modeling [104]. Nevertheless, the transient prediction achieves the required accuracy to determine fuel consumption and gaseous engine-out emissions and the engine model can be implemented into the coupled overall simulation.

4.3.3 Vehicle model

The dynamics of the virtual vehicle are simulated by a mechanical model including variable parameters to adjust the individual components of the powertrain and the vehicle characteristics. In the coupled simulation environment, the vehicle speed has to correlate to the mechanical engine output. In this work, parts of the powertrain are not validated on their own, but the whole vehicle model is verified as a complete system. Therefore, a shaft torque signal (modeled engine torque signal from the ECU) and the engine speed from the measurement of the NEDC are provided as defined input into the mechanical model, as presented in figure 4.12. The simulation result shows that the virtual vehicle follows the measured velocity accurately. The validation confirms the required quality of the vehicle model and thus, it can be applied in the coupled simulation environment.

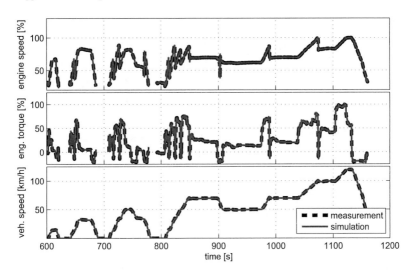

Figure 4.12.: Validation of the vehicle model during the second half of the NEDC

4.3.4 Coupled overall model

After evaluating the quality of each system separately, the behavior of the coupled simulation environment is investigated in the following. The validation of the coupled overall model is challenging, since the intermediate results of the individual sub-models influence the closed-loop calculation directly. First, a global view on the co-simulation matching the simulation results with the corresponding measurement is provided in figure 4.13. As presented on the top, it is important that the driver control can follow the predetermined speed profile and thereby, reproduces a realistic driving behavior. In this example of the NEDC started with a cold engine, the virtual driver can reproduce the measured profile of the acceleration pedal with high accuracy. Furthermore, the signals of the brake and clutch pedal show a comprehensible actuation. As mentioned before, the target speed of the driving cycle has to be met to achieve reasonable predictions about the fuel consumption and emissions. Here, the vehicle speed matches the profile within the legal range of tolerance (+/-2km/h) during the entire NEDC. The performance of the closed-loop overall model and its driver control is confirmed by the small divergences of the cumulated CO_2 emissions compared to the measurement. Again, the high correlation is not only accomplished for the absolute value at the end of the driving cycle, but also for the complete curve progression. The successful

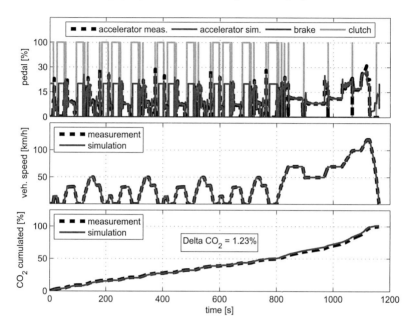

Figure 4.13.: Results of coupled simulation during the NEDC started with a cold engine

performance of the driver control is still confirmed, when looking at the matching of the vehicle speed in detail. In figure 4.14, a period of 50s during the acceleration to 70km/h of the EUDC part within the NEDC is presented. The +/-2km/h limits are based on the measured vehicle speed, since this is the defined target parameter in this simulation methodology. Hence, the boundary lines show the same deflections during the gear shift phases as the basic profile. These little peaks in the vehicle speed demonstrate the complex situation, when the driving is interrupted due to the disconnected clutch. However, the model of the driver control in this work can handle these effects confirming its gear/clutch actuation. Further validation of the driving profile during the second half of the NEDC is shown in figure A.3. It matches the deviations of the simulated vehicle speed to the measurement within the limitation.

Figure 4.15 illustrates the same parameters matched above during the entire RCA started with cold engine. Since this driving cycle has aggressive characteristics, the requirements to the quality of the driver control are even more ambitious. Nevertheless, the simulated acceleration pedal correlates with the measured signal from the test bench. Also, a reproducible operation of the clutch pedal can be observed. Even the dynamic vehicle speed profile of the

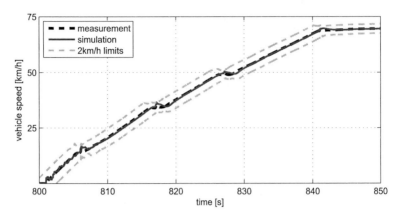

Figure 4.14.: Detailed matching of the vehicle speed during a 50s period of the NEDC started with a cold engine

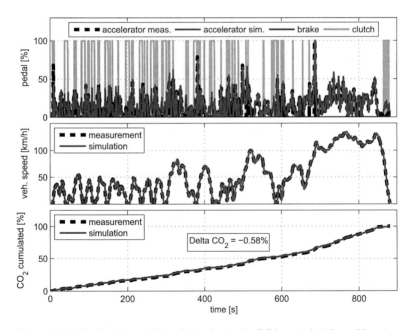

Figure 4.15.: Results of coupled simulation during the RCA started with a cold engine

RCA can be accomplished within the tolerated limits by the driver control. The result of the coupled overall model shows high accuracy in the prediction of the cumulated CO_2 emissions.

Again, the measured vehicle speed can be reproduced in the required quality, even during dynamically periods like the high speed part of the RCA in figure 4.16 and in figure A.4 in the appendix. Up to here, it can be stated that both demonstrated examples confirm the

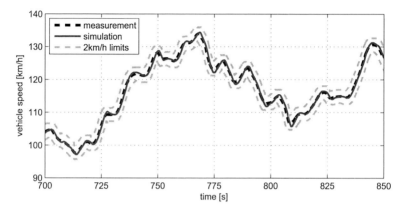

Figure 4.16.: Detailed matching of the vehicle speed during a 150s period of the RCA started with a cold engine

application of this simulation methodology in various driving cycles.

In the coupled overall model, the results of each individual sub-model influence the closed-loop calculation directly and respectively, its quality of prediction. In figure 4.17, the path of the error summation is illustrated. Starting with the driver control, small deviations in the acceleration pedal can generate a considerable impact on the fuel consumption during the driving cycle. According to the approved tolerance range, different driver behavior can result in divergences of the CO_2 emissions, while the target profile of vehicle speed is still observed within the legal limitation. In the ECU model, the modules are simplified and several functions are compensated by substitutes. This can result in some variances in the actuation parameters for the combustion engine. Notable effects on the simulation quality are induced by the engine model. Investigations in this work show a relevant dependence on the gas exchange calculation to the fuel consumption. Little deviations of the air mass trapped in the cylinder can lead to a significant difference in the fuel mass injected. This influences the combustion prediction in the entrainment model, consequently. Thus, the shaft torque is provided by the engine performance. There is a strong interdependency between the engine and vehicle model. Dependent on the engine torque, the wheel torque is determined and accordingly, the vehicle speed. On the other side, the transmission ratios of the powertrain define the engine speed. In order to close the loop, the actual vehicle speed is matched to the target profile by the driver control.

The strong interdependency between the individual sub-models due to the closed-loop simulation makes a separated validation difficult. For example:

- The calculation in the ECU model is mainly based on the position of the accelerator pedal from the driver control, but sensed parameters defining engine and vehicle condition can influence the actuation output significantly.

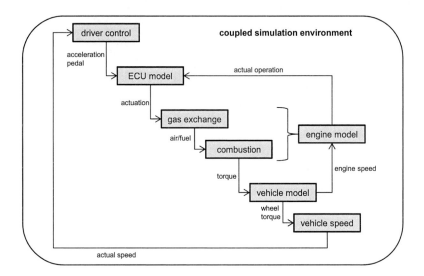

Figure 4.17.: Summation of error in the coupled overall model

- The simulation of the engine performance is a complex system on its own. Thus, the different effects of gas exchange and combustion cannot be investigated separately in the coupled environment.

- Divergences in the engine torque cannot be referred explicitly or distributed to errors within the engine model, deviations in the actuation parameters, and differing vehicle behavior.

- Discrepancies in the vehicle speed due to the summation of the errors in the calculation path are compensated by the driver control resulting in changed input information into the ECU model.

Therefore, several parameters of the sub-models are presented in the following to analyze the performance quality of each component in the coupled system.

The validation of the actuation parameters calculated by the ECU model during the coupled simulation of the NEDC started with cold engine is illustrated in figure 4.18. The quality of the results is comparable to the stand-alone simulation of the ECU model (4.5). A good reproduction of the measurement can be achieved in valve lift and timing. Some deviations can be detected at the spark timing due to peaks at gear shifts. Nevertheless, the profile and range during the driving cycle are in good agreement with the measured values. In summary, the behavior of the ECU model in the coupled overall simulation is similar to the individual validation. It confirms its application and a realistic actuation of the engine model.

In figure 4.19, significant parameters of the engine model during the cold started NEDC are presented. On the top, the engine speed of the coupled simulation follows the profile of the measurement accurately. Only at the idle speed during the cold start period, small deviations can be detected. A good reproduction of the shaft torque shows the comparison

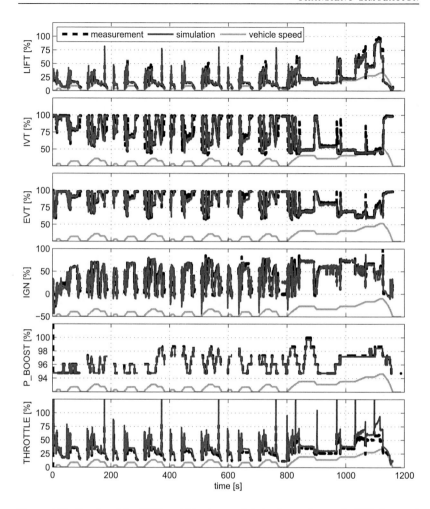

Figure 4.18.: Results of the ECU model in the coupled simulation during the NEDC started with a cold engine

between simulation and measurement in the middle plot. Some discrepancies can be observed at the peak torques during gear shifts. This does not effect the vehicle speed, since the clutch is disengaged in this moment. A small impact on the air flow can be noticed before periods of coasting. Nevertheless, the matching of the air flow overall confirms the good quality in the prediction of fuel consumption.

Comparable results are achieved during the coupled simulation of the RCA started with a cold engine. The matching of the parameters of the engine model show a high congruency

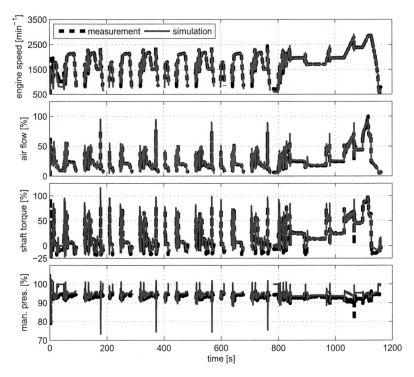

Figure 4.19.: Results of the engine model in the coupled simulation during the NEDC started with a cold engine

to the corresponding measurements, as illustrated in figure A.6. Thus, the demands on the performance of the individual sub-models under various conditions in different driving cycles are also accomplished in the coupled overall simulation environment.

4.3.5 *Emission models*

Beside the fuel consumption, the formation of engine-out emissions in driving cycles can be investigated by this simulation approach. Therefore, the external models described in chapter 3.6 are implemented into the post-processing. Using information gained by the detailed engine model, the engine-out emissions can be determined with respect to the gas exchange and combustion processes in the cylinder. In the following, the validation of the individual models is demonstrated. Remarkable correlations of each pollutant component and the range of application of the sub-models are considered by varied performances of the driving cycles (according to 4.1).

NO$_x$ and CO model

The validation of the engine-out NO$_x$ and CO emissions is achieved by applying the various calibrations due to gas exchange and air-fuel ratio. The amount of engine-out NO$_x$ emissions can be affected significantly by the combustion of residual gas and thus, it is influenced by the calibration strategy for the gas exchange. Since the overlap of valve timing is reduced significantly in the first modification of valve control (var.1), the residual gas ratio is decreased (see section 4.3.2) and subsequently, the forming of NO$_x$ rises remarkably. In figure 4.20, the relative difference of the cumulated engine-out NO$_x$ emissions based on the reference calibration is illustrated during the entire NEDC. Additionally, the resulting profiles of both, simulation and measurement are compared. According to the changed valve actuation, an increase of about 70% can be detected in the measured NO$_x$ emissions. This adverse effect can be reproduced by the emission model. Although the final absolute value of the cumulated divergence is not reached exactly, the characteristics can be replicated during the complete NEDC in good quality and the general tendency is depicted accurately.

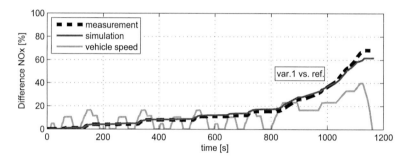

Figure 4.20.: Validation of engine-out NO$_x$ emissions due to variations in calibration strategies

A good reproduction of the measured NO$_x$ emissions is also accomplished in the coupled overall simulation environment. In figure 4.21, the results during the entire NEDC based on the calibration modification var.1 are presented. On the top, the transient emissions referred to the maximum measured value are illustrated. The simulation values do not fit accurately to the data of the measurement, however, the overall profile matches the tendencies very satisfactory. This is confirmed by the good prediction of the cumulated NO$_x$ emissions including the final absolute value. Comparable results are achieved for the reference calibration (ref.) during the NEDC, as well.

The influence of changing the target value of the air fuel ratio to the forming of engine-out CO emissions is presented in figure 4.22. Similar to the validation of the NO$_x$ emissions, the relative differences of the cumulated profiles based on the reference calibration with stoichiometric mixture are illustrated for measurement and simulation during the entire NEDC. Since in the first modification of the injection (var.3) a mixture showing a tendency to fuel-rich (target lambda of 0.95) over the entire driving cycle is burned, the formation of CO is increasing remarkably. Here, the measured emissions are more than doubled referred to the reference. In contrast, the output of CO is reduced at lean combustion (var.4 with a target lambda of 1.05) as expected according to the fundamentals in chapter 1.3. At both variations, the simulated results can reproduce the profiles of the cumulated deviations of the measurements.

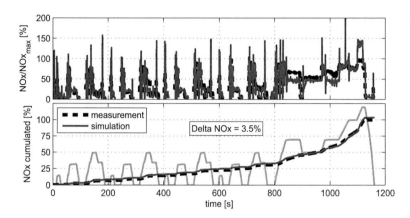

Figure 4.21.: Validation of engine-out NO_x emissions in the coupled overall model during the NEDC with modified valve actuation (var.1)

Even, the absolute amounts of divergency at the end of the NEDC are accomplished with high accuracy.

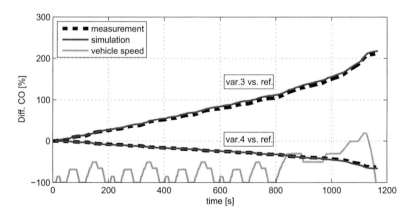

Figure 4.22.: Validation of engine-out CO emissions due to variations in calibration strategies

In figure 4.23, the cumulated CO emissions based on the coupled overall simulation with a fuel rich mixture (var.3) are illustrated during the different driving cycles, NEDC and RCA. Here, the profiles are referred to the measured end value of the NEDC. A slight overestimation during the entire NEDC can be detected comparable to the results based on the input of the engine stand-alone simulation. A similar trend is observable in the RCA. However, the constant overrating of both cumulated profiles remains within a deviation of 2% at the end confirming the satisfying matching to the measurements. This conformity in prediction can be reproduced by the comparison between var.4 and ref. in the same way. Since the behavior of the results is comparable to the calculations with the autonomous engine model, the closed-loop simulation does not effect the quality of predicted CO emissions.

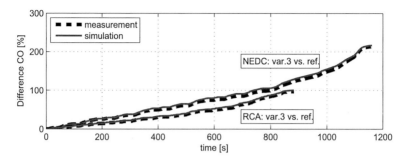

Figure 4.23.: Validation of engine-out CO emissions in the coupled overall model during NEDC and RCA

The successful validation of the engine-out NO_x and CO emission model confirms its application in the coupled simulation environment to determine effects of different calibration strategies during driving cycles. Although the prediction of absolute NO_x emissions is not that accurate than the CO formation, the tendencies can be identified properly.

HC model

Initially, the transient results of the HC model are validated in the NEDC started with warmed-up engine. Comparable to the calculation of steady-state operation, only the mechanisms of piston crevice and oil layer are considered. In the lower half of figure 4.24, a comparison of the cumulated HC engine-out emissions between simulation and measurement is shown. During the first part of the NEDC, the calculated values are continuously too low. But in the extra-urban part, an excessive prediction occurs and balances the undervalued trend. This results in a very good determination of the absolute value at the end of the driving cycle. However, the different observed trends indicate the necessity of further investigations.

The characteristics of the cumulated amount of HC emissions can be analyzed with the help of the instantaneous profile in more detail. This is illustrated in the plot above in figure 4.24. During constant driving periods, the level of unburned fuel masses cannot be reached completely, but the peaks of the acceleration phases are reproduced very well. The underestimation of HC engine-out emissions is most likely due to the modeling of only two formation effects. Further mechanisms like flame quenching and deposits can result in additional emitting of unburned fuel. The high amounts of HC at the second part of the NEDC are mainly based on the overcharge of HC emissions at part load operation known from the steady-state validation (figure 3.28). Additionally effects can be predicated on the overrated fuel consumption from the gas exchange simulation in the engine model during this period.

For the purpose of investigating HC emissions, driving cycles started with a cold engine are of interest. During the period of engine warm-up, the emitting of unburned fuel can increase up to 67% with low engine temperatures [158]. In the lower plot of figure 4.25, the cumulated

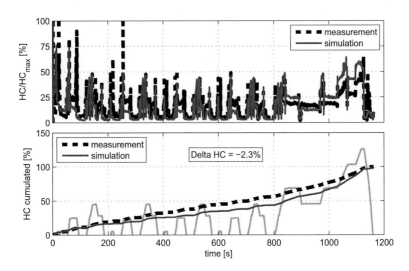

Figure 4.24.: Results of HC engine-out emission in the NEDC warmed-up started

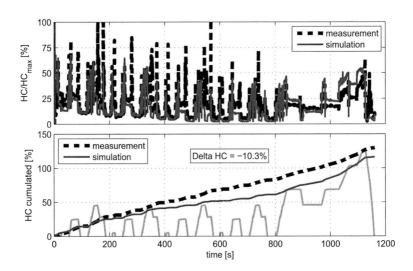

Figure 4.25.: Results of HC engine-out emission in the NEDC cold started

HC engine-out emissions based on the simulation of the cold started NEDC including catalyst heating are compared to the corresponding measurement. Here, the values are scaled relative to the measured HC emissions of the driving cycle performed with warmed-up engine. In this calculation, the formation mechanism of wall quenching is considered as a function of engine temperature (see chapter 3.6.2). Quite good quality of the simulation results is achieved at the beginning of the driving cycle during the period of catalyst heating. From then on, a similar behavior to the warm started driving cycle is observable. However, the deviation of the cumulated characteristic here increases more due to frequent peaks during acceleration and deceleration periods. As shown in the upper half of figure 4.25, the simulation cannot reproduce these high masses of HC emissions exactly. In the extra-urban part, the engine is already warmed-up. Hence, the emitting of unburned fuel is similar to the warm started NEDC, resulting at the end in an undervalued absolute amount of cumulated HC formation about -10%.

Another relevant investigation for the formation of HC engine-out emissions is the quality of the model due to changes of the air-fuel ratio. For this purpose, the NEDC with a constant fuel rich mixture (target value of 0.95 for the air-fuel ratio) and warmed-up engine (var.3) is considered. The results are matched to the cumulated HC engine-out emissions of the measured driving cycle in figure 4.26. Again, a similar behavior of the simulation compared to the transient performances above is identifiable. The relative deviation of the final absolute amount is about -6% and confirms, in combination with the other presented results, the implementation of the HC sub-model in the simulation environment to support engine calibration processes.

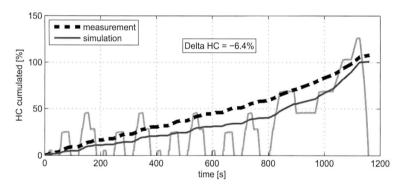

Figure 4.26.: Results of HC engine-out emission in the NEDC with constant fuel rich mixture and warmed-up engine (var.3)

The prediction of HC emissions is also investigated in more dynamic driving situations. Figure 4.27 presents the cumulated profiles of HC engine-out emissions during the RCA for the three variations already discussed above. All cumulated HC emissions are referred to the measured RCA started with a warm engine. The result of the simulation performed with a warmed-up engine shows a good matching with the corresponding measurement. Comparable to the NEDC, a little underestimating can be detected in the first part of the driving cycle and excessive prediction occurs at higher engine speeds and loads. Starting with a cold engine, a good quality is achieved during the period of catalyst heating. In the rest of the RCA, a similar behavior to the NEDC can be observed. The profile of the variation of the air-fuel ratio can be reproduced with high accordance. The end value can be determined even better than in the NEDC, since the RCA is driven with a target lambda of only 0.97. It can be stated that the similar results of different transient conditions confirm the application of the HC model in driving cycle simulations.

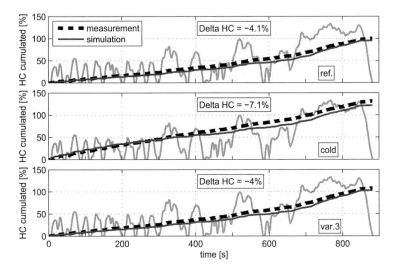

Figure 4.27.: Results of the cumulated HC engine-out emissions compared for different performances of the RCA

The quality of the HC model depends on the input parameters calculated by the engine simulation. In order to evaluate the prediction of HC engine-out emissions in the coupled overall simulation environment, several results of the NEDC and RCA based on the closed-loop engine calculation are presented in figure 4.28. All profiles are referred to the cumulated end value of the measured NEDC performed with warmed-up engine. During both driving cycles, the same underestimating in the prediction of HC emissions as the calculation based on the results of the engine stand-alone simulation can be detected. The validation confirms a similar behavior of the engine model independent from effects of the closed-loop system.

4.3.6 *Summary*

After the successful completion of the validation process, it can be stated that the applicability of the presented methodology for the virtual calibration is confirmed. Before the sub-models can be coupled to the overall simulation environment, each system has to be optimized and matched individually. In the ECU model, the focus is on reproducing equivalent quality in the calculation results although the complexity of the functions is reduced. The achieved quality of the resulting actuation parameters confirms its modeling depth. Investigations in the engine model show a strong influence on the prediction of fuel consumption by the gas exchange simulation. Due to analysis of various valve actuation strategies, parameters defining the flow characteristics at throttle and valves can be adapted in the model. The calibration results in a high quality of the gas exchange, but there is still a potential left for modeling issues. Therefore, further researches and measurements obtaining more information about the air

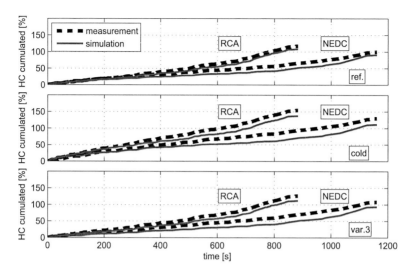

Figure 4.28.: Results of the cumulated HC engine-out emissions in the coupled simulation compared for variations of NEDC and RCA

flow characteristics of engines with full variable valve timing and turbocharging are required. The implementation of the vehicle model as a simplified mechanical, nearly inelastic 1D-multibody system conforms to its requirements. If more detailed specifications are requested, it can be easily expanded and several components can be added or substituted. Based on the high quality of the individual sub-models, the coupled simulation environment achieves satisfactory results in various driving cycles. Here, the validation shows some difficulties, since deviations of the sub-systems cannot be analyzed separately. The path of a possible error summation in the closed-loop simulation is presented and exemplified.

According to various measurements of different calibration strategies in valve actuation and gas mixture, the transient performance of the engine-out emission models is validated. The HC model is compared successfully to cold and warm engine conditions and a fuel rich air-fuel ratio. Nevertheless, the results indicate potential improvements due to further formation mechanisms like flame quenching and liquid fuel. Therefore, additional specific measurements are required and the level of detail in the modeling has to be increased. A similar quality is accomplished in the prediction of NO_x emissions. However, the dependency on the residual gas ratio can be reproduced correctly. High accuracy is achieved in the simulation of the engine-out CO emissions due to various air-fuel ratios.

Summing up, the simulation methodology accomplishes satisfactory quality under various conditions and different transient situations. After proofing the concept by an extensive validation, the results confirm its application in further investigations.

5

APPLICATION EXAMPLES

The results of the validation confirmed the application of the presented simulation methodology in driving cycles to analyze fuel consumption and engine-out emissions. In this chapter, further investigations enabled by the coupled simulation environment are introduced. Changes in the legislation regarding emission regulations are envisioned in the near future. In the first section (5.1), the new test procedure is described and simulation results are evaluated. Another important effect considering fuel consumption and emissions is the profile of the air-fuel ratio during a driving cycle. Therefore, its impact is analyzed and a modeling approach is demonstrated in section 5.2. A detailed look into transient periods is provided in section 5.3 showing one advantage of the simulation methodology. In the last section (5.4), the methodology of determining tailpipe emissions is implemented into transient calculation of driving cycles.

5.1 WORLDWIDE HARMONIZED LIGHT-DUTY TEST PROCEDURE (WLTP)

In the last years, the NEDC was often criticized not to represent real-life driving. Long constant speed cruises, many idling events and only low engine loads are not common with transient accelerations in world-wide practice. Its potential successor, a new Worldwide harmonized Light-Duty Test Procedure (WLTP), is currently in development (Bielaczyc et al. [5]). The aim is to design a common set of driving cycles for all passenger cars, independently from their characteristics. While the speed profile is nearly fixed, the definition of ambient conditions regarding the dynamometer test and car settings like optional equipment, tires, and vehicle payload is still in progress. A first implementation of the WLTP in Europe is expected for 2017 (Mock et al. [116]).

Classification and specifications

The test procedure is basically divided into three different driving cycles depending on the vehicle class (UNECE Global Technical Regulation No. 15 [168]). This class is defined by the power to unladen weight ratio of the vehicle in W/kg:

- Class 1: low power vehicles with Power/Weight $<= 22$ W/kg

- Class 2: vehicles with $22 <$ Power/Weight $<= 34$ W/kg

- Class 3: high power vehicles with Power/Weight > 34 W/kg

Class 3 is additional separated into two subclasses, a and b, belonging to the potential maximum speed either below ($<$) or greater than or equal ($\geq$) to 120km/h (UNECE Global

Technical Regulation No. 15 [168]). In this work, only Class 3b is relevant, because the investigated vehicle used throughout this work has a power-weight ratio of 141kW/1250kg and a maximum speed of 235km/h.

The speed profile (shown in figure 5.1) is designed to represent real vehicle operation world-wide on urban and extra-urban roads, motorways, and freeways. The WLTC (Worldwide harmonized Light-Duty Test Cycle) for a Class 3 vehicle consist of four parts subdivided into the vehicle speed periods Low, Medium, High, and Extra High speed. Compared to the NEDC, the standing phases are reduced and more dynamic parts with higher gradients in acceleration are included.

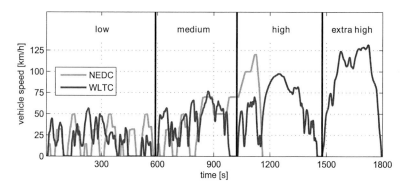

Figure 5.1.: Vehicle speed profile of the WLTC

One substantial objection against the NEDC is the definition of fixed gear shift points for vehicles with manual transmissions. In the last years, the number of forward gears increased and there are vehicles with 4 up to 7 gears available. However, the gear selection and shift points in the WLTP are determined for every vehicle individually by an approach according to Economic Commission for Europe [28] described in the following. A transient profile of gear changes can be generated by functions of gear ratios, engine speed, and acceleration/power demand. Additionally, specific vehicle parameters like test mass, maximum engine power, and idling speed of the combustion engine are taken into account.

The required power $P_{required,j}$ in kW to attain the acceleration and maximum speed of the driving cycle is calculated for every second j of the cycle trace by

$$P_{required,j} = \frac{f_0 v_j + f_1 v_j^2 + f_2 v_j^3}{3600} + \frac{k_r \dot{v}_j v_j m_{veh}}{3600} \quad (5.1)$$

with f as factors of the road load. The coefficient (f_0 in N) is defined as force, parameter $f_1 \left[\frac{N}{km/h}\right]$ is dependent on velocity and $f_2 \left[\frac{N}{(km/h)^2}\right]$ is based on the square of velocity. v_j [km/h] is the vehicle speed, while $\dot{v}_j$ [m/s^2] is the acceleration at second j

$$\dot{v}_j = \frac{v_{j+1} - v_j}{3.6} \quad (5.2)$$

The vehicle test mass is described by m_{veh} [kg] and kr is a factor taking into account the inertial resistances of the drivetrain during acceleration. Converted into non-dimensional form, a normalized full load power curve $(P_{norm,fullload})$ defining the percentage of rated power available at a specific engine speed rpm_{norm} is established:

$$P_{norm,full\,load} = \frac{P_{fullload}\left(rpm_{norm}\right)}{P_{max}} \tag{5.3}$$

The operation range of the engine is also normalized to the rated engine speed at maximum power $(rpm_{P_{max}})$ minus idling speed (rpm_{idle}):

$$rpm_{norm} = \frac{rpm - rpm_{idle}}{rpm_{P_{max}} - rpm_{idle}} \tag{5.4}$$

Based on this normalized full load power curve, the available power for each possible gear number g and each vehicle speed value of the cycle profile v_j is expressed by

$$P_{available_{g,j}} = P_{norm,full\,load}\left(rpm_{norm_{g,j}}\right) \cdot P_{max} \cdot SM \tag{5.5}$$

The difference between the power curve based on stationary full load and the real power available during transient conditions is incorporated by the safety margin SM [-]. For each gear the normalized engine speed is calculated using equation 5.4:

$$rpm_{norm_{g,j}} = \frac{ratio_g \cdot v_j - rpm_{idle}}{rpm_{P_{max}} - rpm_{idle}} \tag{5.6}$$

where $ratio_g$ is the the global ratio of the powertrain obtained by dividing rpm $\left[\text{min}^{-1}\right]$ by v $\left[\frac{\text{km}}{\text{h}}\right]$ for each specific gear g.

At each second of the speed profile j, possible gears are limited by the condition providing enough power:

$$P_{available_{g,j}} \geq P_{required_j} \tag{5.7}$$

Following additional requirements have to be considered during determination of permitted gears. Only resulting engine speeds that fulfill

$$rpm_{min} \leq rpm_{g,j} \leq rpm_{max} \tag{5.8}$$

are allowed while driving the cycle trace at v_j. The limits are dependent on the gear number and engine characteristics. In order to represent practical use and fuel efficient driving style, several additional requirements have to be observed. Exemplarily, frequent gear changes occurring in less than 5 seconds have to be filtered out and gears used during accelerations and decelerations have to be retained for a period of at least three seconds.

Finally, the gear profile determined for the drive-train/engine assembly of the selected vehicle during the WLTC is shown in figure 5.2. Further input data required to simulate this new driving cycle is summarized in table 5.1. Since no measurement of exactly the vehicle setting investigated in this work is available, the parameters are estimated from a test of a forerunner engine in combination with a different vehicle type.

Simulation results

In the following, the simulation results of the coupled overall model are presented and discussed in comparison to the NEDC. First, the simulated vehicle speed is matched to the prescribed target driving profile. A detailed look at a section in the middle of the WLTC

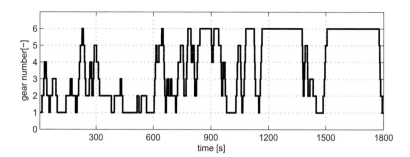

Figure 5.2.: Determined gear profile during the WLTC

Required input	Implementation	Source
gear number	profile as function of time	calculation
demanded vehicle speed	profile as function of time	regulations
clutch actuation	profile as function of time	dependent on shifting/veh. speed
air-fuel ratio	profile as function of time	measurement of related engine
oil temperature	profile as function of time	measurement of related engine
cooling water temperature	profile as function of time	measurement of related engine
torque loss by auxiliaries	profile as function of time	measurement of related engine
ambient conditions	constant values	measurement of related engine

Table 5.1.: Summarizing of required input data to simulate the WLTP.

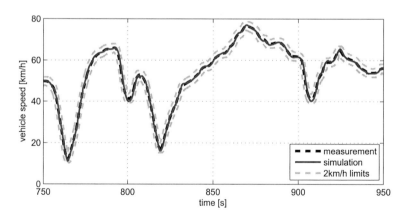

Figure 5.3.: Detailed matching of the vehicle speed during a 200s period of the WLTC

with high gradients in acceleration is provided in figure 5.3. Further validation of the speed control during the low speed part is presented in the appendix (figure A.5). In Figure 5.4, main parameters of the modules driver control and combustion engine are shown. On the top, the vehicle achieves the required target speed over the entire WLTC confirming the defined gear profile at the same time. The driver control shows a smooth behavior of acceleration and brake pedals and no abnormal effects can be detected.

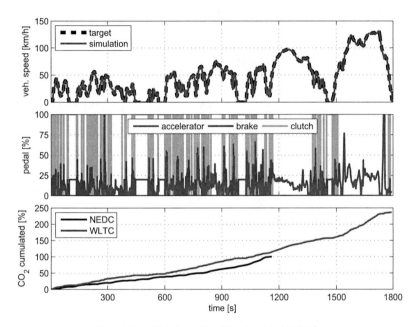

Figure 5.4.: Global results of the simulated WLTC

Compared to the NEDC, more than 236% of the total amount of CO_2 are emitted at the end of the WLTC. Related to the driven distance, it results in an increase of 12% of CO_2 in g/km. In the literature, the effect of the changed test procedure on the fuel consumption is discussed controversial. The automobile press prognosticate increased values up to 25% depending on the weight class and motorization [129, 151]. According to an investigation of the ADAC (Allgemeiner Deutscher Automobil-Club e.V.) with 352 measured passenger cars (Mock et al. [116]), a wide spread of -15/+12% (including error tolerances) in the difference of fuel consumption is detected depending on petrol or diesel powertrain. This is similar to measurements by Marotta et al. [101] with varying conditions of vehicle test mass. A study by ICCT and Ricardo Inc. [116] based on simulations with a physics-based vehicle and drivetrain system model predict 14% higher CO_2 emissions in the WLTC for a small car with direct injection, start-stop technology, and automatic transmission. This demonstrates the difficulty to make a reasonable forecast and confirms the development of simulation methodologies. It must be pointed out that the simulation results in this work are only valid for the made assumptions according to table 5.1. Furthermore, fuel consumption could be improved by optimizing the gear ratios to the new speed profile and by adapting the engine calibration.

In figure 5.5, several main parameters of the engine model are summarized. Compared to the NEDC, the engine speed is in a similar range and meets the requirement regarding the limits rpm_{min} and rpm_{max} while driving. Air flow and shaft torque (referred to the maximum value of the NEDC) show higher values in the arithmetic average, especially in the high and extra high part. This confirms the demand on increased load and results in more fuel consumption. The level of the manifold pressure is nearly the same as in the NEDC, but the profile includes more dynamics. Even the required load is increased, the power of the investigated engine is still sufficient and additional boost by the turbocharger is rarely necessary.

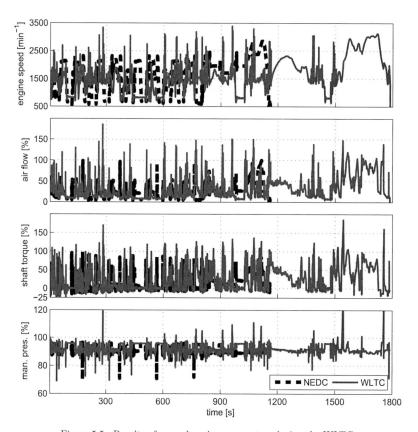

Figure 5.5.: Results of several engine parameters during the WLTC

Results of the actuation parameters calculated by the ECU model during the WLTC are presented in figure 5.6. All profiles are referred to the corresponding maximum value of the NEDC simulation. The increased opening of the intake valve verifies the higher air flow rates. The short peaks in the periods of low and medium vehicle speeds show potentially improvement of the starts of acceleration and shifting phases in the simulation. The values of the valve timing and spark ignition are in the same range as in the NEDC and demonstrate a reasonable behavior. The actuation of the throttle is very smooth (except of shifting) and comparable to the NEDC. Summarizing, the results validate the application of the ECU model in further driving cycles like the WLTC.

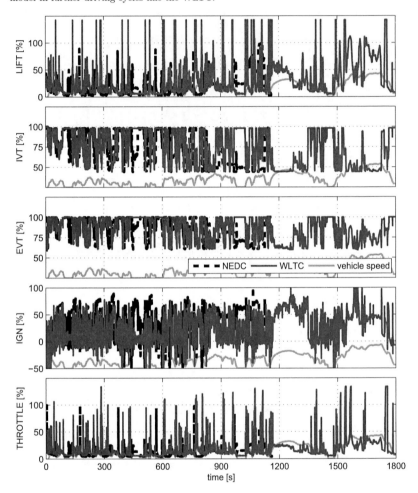

Figure 5.6.: Results of the actuation parameters calculated by the ECU during the WLTC

5.2 SENSITIVITY OF AIR-FUEL RATIO

When analyzing fuel consumption of combustion engines, the air-fuel ratio has a considerable influence on the injection. Oscillations in the lambda signal can occur due to inadequate control behavior of the ECU, inaccuracies in the load detection (sensor) and injection (actuation), and requirements of stimulating the conversion rate of the TWC (e.g. oxygen storage).

However, it is difficult to reproduce a realistic signal of the air-fuel ratio in virtual investigations. Until now, in this simulation methodology the measured profile from the engine test bench is used during validation of the models in chapter 4. Since the overall coupled approach is designated to make predictions about prospective concepts, the calculation has to be as independent from measurements as possible. Therefore, an investigation of the sensitivity in fuel consumption based on the air-fuel ratio with the aim to generate a virtual signal is presented in the following.

Constant profile of the air-fuel ratio

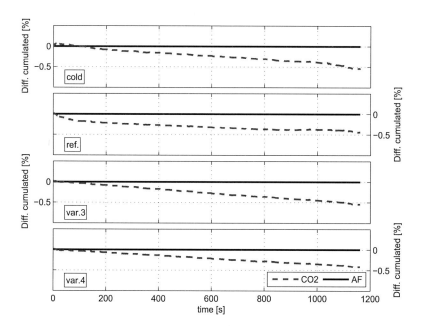

Figure 5.7.: Matching of engine stand-alone simulations provided with constant and measured profile of the lambda signal

In this sub-section, the influence of the oscillations in the lambda signal during a driving cycle is investigated. Therefore, simulations with a constant profile of the air-fuel ratio are performed. In the first comparison, results of several calibration variations (from chapter 4.1) simulated with the engine stand-alone model are demonstrated in figure 5.7. It shows the relative deviation of cumulated fuel consumption (CO2) and air flow (AF) between the simulations with constant and measured profile of the lambda signal during the NEDC. Overall, a clear tendency of the cumulated CO_2 can be observed. The constant value of the air-fuel ratio without any oscillations results in a lower fuel consumption of about 0.5% at the end of the cycle. No difference can be detected in the air flow, since the actuation of the gas exchange remains constant in both simulations. Thus, less fuel is injected and the power output of the engine is reduced. Eventually, this can have an impact on the vehicle speed.

In figure 5.8, the simulations with predetermined constant value of 1.0 and the measured signal of the lambda profile are performed by the coupled overall model ensuring the vehicle speed profile of the driving cycle is still achieved by means of the virtual driver control. During the NEDC with cold started engine and activated catalyst heating, the relative differences of CO_2 and air flow show a comparable behavior to the results of the engine stand-alone comparison. Nearly no deviation can be detected in the matching of the cumulated engine torque (TQ). It means, the reduced fuel injection over the whole driving cycle by the constant air-fuel ratio does not effect the power output of the engine. This is confirmed by the conformity of the cumulated vehicle speed (VS) between both simulations and the similar behavior of the driver control (presented by the cumulated profile of the accelerator pedal (AP)). Related results are observed during the NEDC started with a warmed-up engine. A reduction of the air flow and fuel consumption is noticeable, but it seems to have no impact on the engine torque and vehicle speed (the relative difference of both parameters range around zero). Certainly, the constant profile of the air-fuel ratio is not realistic and thus, the utilization of these simulation results is restricted.

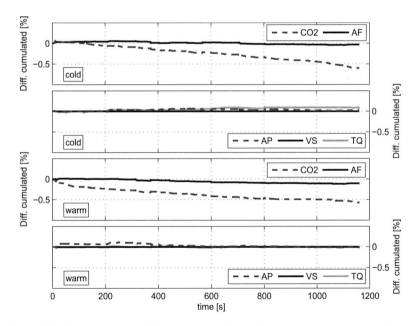

Figure 5.8.: Matching of coupled simulations provided with constant and measured profile of the lambda signal

Generated profile of the air-fuel ratio

In the next step, the signal of the air-fuel ratio is modeled trying to reproduce the influence of its oscillations. Since in the ECU model the control of the air-fuel ratio is not modeled yet, dependencies of this parameter have to be identified and implemented. The original module from the ECU is not adopted directly, since the adaption of the lambda signal consists of a complex artificial neural network. For the purpose of an alternative model, a analogous function of the frequency and amplitude according to the dynamics of air flow is observed in the measured profiles of the air-fuel ratio. The schematic structure of the simplified calculation is illustrated in figure 5.9. It consists of two sinus waves, one for the basis frequency and the other to stimulate its amplitude. The frequency and peaks (algebraic sign and absolute level) of the amplitudes are dependent on the change in the air flow using its filtered derivative and intensifying the signal by multiplication. After filtering the amplitude signal again, it is added to the basis frequency. At the end, the lambda signal is set to oscillate around the value one (base line) and limited to avoid numerical outliers.

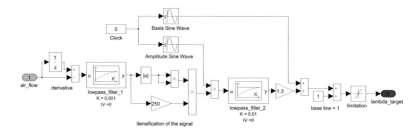

Figure 5.9.: Model of the lambda signal

The resulting virtual profile of the lambda signal is compared with a measured period in figure 5.10. It shows a example with a length of 300s including sections of acceleration and steady driving. The simulation does not exactly emulate the characteristics produced by the real engine, but significant changes in the oscillation are correlating to the different situations of the speed profile. The amplitudes are small during steady-state conditions, whereas the frequency and peaks increase in the periods of acceleration. Thus, the requirement for the modeled approach to reproduce the characteristics of the measured profile and to oscillate within the appropriate scope of amplitude and frequency is accomplished.

In figure 5.11, the relative deviations between the simulations with generated and measured profile of the lambda signal are illustrated. In both variations of the NEDC, started with cold and warmed-up engine, the fuel consumption matches very good. Some discrepancies can be detected in the air flow, but still in a small range. They can be explained by the differences of the cumulated signal of the accelerator pedal. Here, the influence of the driver control seems to be stronger than the perfect matching of the lambda oscillations. It adjusts the reactions in the closed-loop system continuously, since the actuation of the accelerator pedal is a direct reaction of the actual vehicle speed resulting from the power output of the engine and thus, the injected fuel mass. There is evidence to suggest that the modeled signal is too lean during the acceleration periods corrected by higher accelerator pedal positions. In conclusion, the vehicle speed profile can be obtained based on only small deviations in the

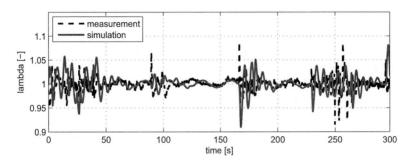

Figure 5.10.: Validation of the generated lambda signal

driver actuation and the fuel consumption is still matched on the same time. The presented results confirm the application of the simplified lambda signal generation and offer advanced investigations of fuel consumption or emissions in the future due to air-fuel ratio effects.

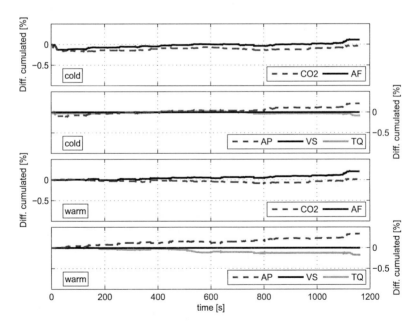

Figure 5.11.: Matching of coupled simulations provided with generated and measured profile of the lambda signal

5.3 APPLICATIONS OF CRANK ANGLE BASED ANALYSES

Crank angle based analyses of engine-relevant quantities are not possible in typical fuel consumption simulations with simple empirical, map-based engine models. However, the access to this level of information can give valuable insights for the design and calibration of an engine. In this section, an extract of examples with focus on scaling up is described. In order to demonstrate the flexible application of different time ranges exemplarily, two crank angle based analyses are shown in figure 5.12. On the left side, combustion relevant quantities like residual gas concentration, cylinder pressure and turbulent burning velocity are plotted within the extra-urban part of the NEDC. Besides the cylinder pressure, the measuring of parameters characterizing the combustion process is often complicated and costly. Usually, serial engines are not equipped with required sensor technologies. This is similarly to vehicles, typically used on roller dynamometers to monitor fuel consumption. Here, it is actually more complex to install relevant instrumentation. Even though specialized test equipment is available, there are still parameters left that are not measurable directly. The turbulent burning velocity can only be determined by simulations, for example.

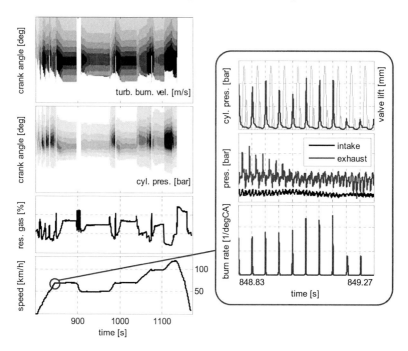

Figure 5.12.: Example of crank angle based analyses

In a very short time span of approximately 0.5s, the cylinder pressure together with valve lift, intake/exhaust manifold pressure and normalized burn rate are depicted for one cylinder of the engine on the right side of figure 5.12. This example shows a possible analysis of transient engine operation in a specific time period of a driving cycle. Here, the change of acceleration to steady driving at the beginning of the extra-urban part is depicted. This is an interesting situation for the calibration, since the fast change in engine load involves a highly dynamic control of the ECU. After attaining the target speed, low power is required and the actuation of the engine has to be adapted. This is shown in the top plot by the decreasing valve lift, exemplarily. Also identifiable is the remarkable diminishing of the cylinder pressure as immediate reaction by the combustion. Last one is represented by the burn rate in the bottom plot. Not only parameters in the cylinder response, also the entire path of air flow is affected by the changed actuation of the ECU adjusted to the driver's request. One result is observable in the pressure oscillations in the intake and exhaust manifolds.

The advantage of detailed information about engine working cycles provided by this simulation methodology can be applied for several comprehensive investigations. One application is already mentioned and presented in the analysis of gas exchange effects in chapter 4.3.2. With the help of focusing in individual working cycles, the airflow streaming in and out of the cylinder is analyzed. Influences of the valve actuation are detected and the resulting pressure profiles in intake and exhaust manifold can be evaluated. This information can provide useful support of calibrating valve timing.

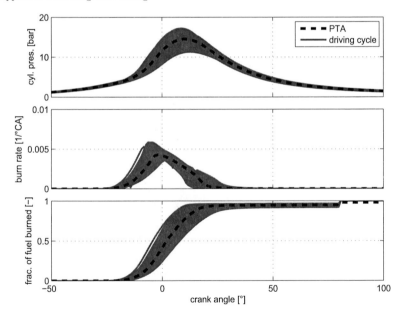

Figure 5.13.: Matching of several working cycles during the driving cycle simulation with results of the PTA

Further application is demonstrated in the figure 5.13. It shows a comparison of about 200 working cycles during a steady drive period of 15s in the NEDC. These are matched with the result of the PTA of an equivalent steady-state operation point for the combustion parameters cylinder pressure, burn rate, and fraction of fuel burned. The different working cycles during the transient simulation move observably around the PTA profiles. The oscillations are a result of the dynamic system including driver, ECU, and engine. In contrast to the one working cycle of the steady-state situation, the engine speed and load is differing within a tiny range in the closed-loop environment. This is induced by several impacts affecting each other. For example, the driver does not keep the accelerator pedal in constant position perfectly. The control functions of the ECU are influenced by small dynamics in the signals, as well. Finally, the air path of the engine with its slow reactions provokes continuous adaptions of the ECU.

Further application possibilities to support calibration tasks are zooming in periods of shifting, overrun fuel cut-off, and Tip-In maneuvers. Also, engine-out emissions can be analyzed in detail. In summary, the interplay between engine calibration and thermodynamics can be studied and optimized in any time range, i.e. on demand.

5.4 TRANSIENT SIMULATION OF TAILPIPE EMISSIONS

In this section, the post-processing to determine tailpipe emissions from the steady-state prediction (3.7) is implemented into transient calculation of driving cycles. Therefore, some adaptions executed in the measurements and assumptions are presented. Modifications of the model due to transient conditions are introduced and adjustments are explained. Finally, the simulation results are matched to the measurements and improvements of the model quality are discussed.

Changes in measurements and specifications due to transient effects

In order to investigate transient results of tailpipe emissions, measurements of an entire NEDC were performed on the dynamometer test bench. The powertrain of the vehicle consists of a four-cylinder engine combined with the catalyst from chapter 3.7.2. Loading and aging state of this catalyst are different to the measurements of the light-off characteristics, but were not further specified. However, in this investigation, the parametrization of an aged catalytic converter from the light-off analysis is assumed. The gas phase species of CO, CO_2, NO_x, and O_2 were observed dry, thus requiring an adaption by the water-correction factor (H_2O content). THC is measured wet and can be used directly. The exhaust gas is extracted in front of the catalytic converter and at the outlet of the exhaust system instead of directly behind the catalyst as in the case of the steady-state analysis. The large distance between these two points of extraction results in a significant time lag in the signals depending on system pressure and mass flow rate. Due to this extensive dynamic effect, the evaluation of time-resolved variables is not reasonable possible. The cumulated masses of the pollutants instead are preferred for validation of simulations. A comparison of the cumulated tailpipe emissions, illustrated in figure 5.14 exemplarily, shows their different formation progress. Noticeable is the sequence of the first 100 seconds, where already more than 40% of the total emissions is emitted.

During the analysis of the stationary light-off experiment, an approximately homogeneous temperature distribution in the catalyst can be assumed (as described in section 3.7). Due to calibration effects of the combustion engine at the beginning of a driving cycle, the monolith is heated rapidly with high temperature gradients in both directions, axial and radial. The warm-up behavior of the catalyst considering inlet gas conditions, heat transport, and reaction

129

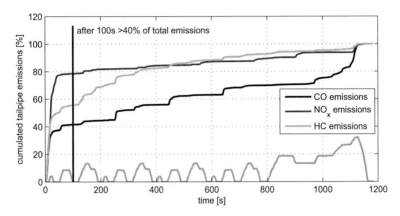

Figure 5.14.: Cumulated tailpipe emissions relative to their total end amount [38]

enthalpies is not implemented in the model, yet. However, a wall temperature can be specified time resolved from the measurement. This enables the definition of an axial temperature profile inside the plug module for each time step.

A radial temperature distribution can be achieved by subdividing the cross-sectional area into smaller circular zones that are estimated at constant temperature. This implies a uniform gas flow centered at the intake of the catalyst. Each zone is calculated by one plug at a specific temperature, representatively. Since every channel is simulated independently without interaction to its environment, the heat transfer in radial direction cannot be reproduced exactly by this simplified approach.

The temperature distribution in both directions, radial and axial, is obtained from a measurement with a modified catalyst containing several thermocouples at different positions. Some temperature profiles during the complete NEDC at various locations inside the monolith are compared in figure 5.15. In axial direction a time lag after the first half along the flow direction can be observed, especially during warm-up of the combustion engine in the urban part of this driving cycle. Analyzing the radial distribution, only slightly deviations can be observed between the temperature profiles at the center of the catalyst and 53% of the cross-sectional radius outwards. However, a reduced temperature level can be detected in the area of the outer ring. More detailed information about the implementation of the wall temperature profiles in the catalyst model is provided in Frommater [38].

Transient model of tailpipe emissions

Compared to the application calculating the light-off behavior, the process steps have to be adapted to simulate driving cycles. The extended simulation process scheme is illustrated in figure 5.16. Especially, the preprocessing of the larger amount and dimension of input parameters has to be implemented. Due to the long period and transient conditions of driving cycles, the size of time-dependent variables increases remarkable. The simulation methodology to calculate the conversion rate during a specific time step consists of a global reaction mechanism, effective pore diffusion, and the plug flow approach, similarly to the

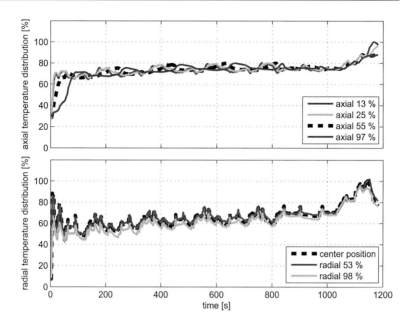

Figure 5.15.: Radial and axial temperature distribution at various locations inside the monolith [38]

steady-state application in chapter 3.7. Since the catalytic converter is no longer considered as isothermal, the temperature distribution inside the monolith has to be implemented. The wall temperature is adopted from the approach of the heat distribution and can be adjusted in both directions, axial and radial. While the temperature in axial direction can be applied from the measurement above directly, the radial change is considered by calculating two channels independently. One represents the center of the catalyst (inside of 75% of its radius), the other is placed in the outer ring (remaining 44% of its whole cross-sectional area).

Each time step is calculated individually considering the driving cycle as quasi-stationary state. This approach is based on the assumption that the reaction rates and mass transfers to the catalysts surface are not affected by any species accumulation phenomena on the surface. However, transient effects are implicated by an additional sub-model calculating the oxygen storage and release (Tsinoglou et al. [165]). After every time step Δt, the mass of each species $m_{k,t}$ at time t is converted by the concentration profile $c_{k,t}$ and the measured mass $m_{k,meas}$ and mass fraction $Y_{k,meas}$ of the particular species k:

$$m_{k,t} = c_{k,t} \frac{M_k}{\sum_{k=1}^{N_G} M_k} \frac{m_{k,meas}}{Y_{k,meas}} A_{rep} \qquad (5.9)$$

The resulting masses are then cumulated by

$$m_{k,cum} = \sum_{t_{start}}^{t_{end}} m_{k,t} \cdot \Delta t. \qquad (5.10)$$

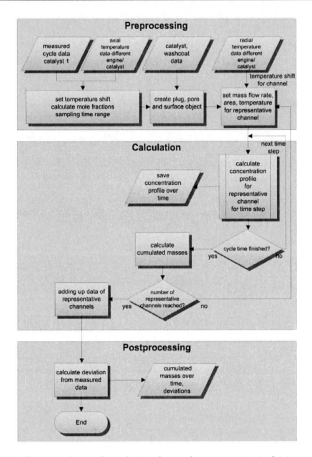

Figure 5.16.: Process scheme of simulating the catalyst conversion in driving cycles [38]

Since only two representative channels are calculated, the cumulated masses of both have to be added up proportional to their respective areas A_{rep} to consider the entire catalyst. In the post-processing, the deviations of the total cumulated masses during the driving cycle are identified.

One essential effect in three-way catalysts during transient operation is the storage and release of oxygen. Theoretically, the ECU ensures a mixture near stoichiometric by the usage of oxygen sensors. Due to transit time of the exhaust gas and the sensor's response delay, the system has some time lag generating oscillations in the air to fuel ratio [64, 84]. In order to compensate the degradation of the conversion rate at air-fuel ratios outside of the ideal range, cerium is added to the washcoat. This reacts with the gas phase species and plays an important role in dynamic oxidation-reduction phenomena (Tsinoglou et al. [165]).

Cerium has the capacity to store oxygen by forming three-valent Ce_2O_3 and four-valent CeO_2, expressed by the oxidation reaction:

$$Ce_2O_3 + 0.5O_2 \rightharpoonup 2CeO_2 \tag{5.11}$$

According to reaction mechanisms cf. Holder [63] (see table A.2), the following equations are added to the oxygen storage model:

$$H_2 + 2CeO_2 \rightleftharpoons H_2O + Ce_2O_3 \tag{5.12}$$
$$CO + 2CeO_2 \rightleftharpoons CO_2 + Ce_2O_3 \tag{5.13}$$

Additional reactions of cerium are reported in literature, but not considered here. Tsinoglou et al. [165] include the oxidation of Ce_2O_3 by NO. The reaction of cerium with hydrocarbons is assigned to be of minor importance (Herz [58]). During fuel-rich mixtures of the exhaust gas, the oxidized cerium provides oxygen to react with CO and H_2. At lean gas conditions though, it can absorb O_2 and reduce NO at the same time (Koltsakis et al. [84]).

The oxygen storage capacity is affected apparently by the cerium loading and dispersion in the washcoat. Its correlation can be expressed as the active site density of cerium Γ_{Ce}. Holder [63] defines its value per reactor volume $[kmol/m^3]$, but it can also determined as function of the active surface area $[kmol/m^2]$ referred to the literature. Based on the surface of one channel, it can be calculated by specifying an oxygen storage capacity OSC $[gO_2/l]$:

$$\Gamma_{Ce} = 2\frac{OSC}{M_{O_2}}\frac{d_{channel}}{4F_{CatGeo}} \tag{5.14}$$

The factor 2 depends on the condition that one O_2 molecule requires two Ce_2O_3 sites (see 5.11). Since no specific information about the washcoat is available, the oxygen storage capacity is assumed to a typical value of $OSC = 2gO_2/l$ (Frommater [38]).

The surface species m (CeO_2 and Ce_3O_3) are balanced by means of the surface coverage θ_m defined by the ratio of occupied sites of species m to total number of sites (Holder et al. [64]). The time-dependent change of the surface coverage that is scaled between 0 and 1 can be expressed by

$$\frac{d\theta_m}{dt} = \frac{1}{\Gamma_{Ce}}\sum_{r=1}^{N_R}\nu_{mr}\omega_r \tag{5.15}$$

with the stoichiometric coefficient of the reaction ν_{mr}. According to the reaction mechanism, the rate expressions ω_r are defined by

$$\omega_{21} = k_{21} \cdot X_{O_2}\theta_{Ce_2O_3}, with \tag{5.16}$$
$$k_{21} = A_{21} \cdot e^{-\left(\frac{E_{21}}{RT}\right)} \tag{5.17}$$

The rate expressions are added to the global reaction mechanism to implement the calculation of the oxygen storage. The solving of the reaction rates requires the coverage state of the channel along the axial direction for each time step. After initialization of $\theta = 1$, a concentration profile is calculated quasi-stationary. In this case, the initial condition represents a catalyst fully loaded with oxygen at the beginning of the driving cycle. Based on these concentrations, the surface coverage of the next time step is determined in a loop by using equation 5.15. The process scheme of the oxygen storage module is illustrated in figure A.7. In order to generate a profile along the axial direction, the length of the channel is divided into several grid points with linear interpolation between them (Koop [86]).

The adjustment of the kinetic parameters and the oxygen storage capacity can be done with the help of laboratory scale experiments described in Holder [63]. Since no measurements with synthesis gas were available for this investigation, the parameters were adopted from Holder [63]. Afterwards, the global reaction mechanism including the oxygen storage equations is matched by a global shift of all parameters to the driving cycle measurements. The detailed process of the optimization is described in Frommater [38].

Simulation results

In figure 5.17, the simulated tailpipe emissions of the main pollutants during the entire NEDC are compared to the measured profiles. All results are scaled to the maximum cumulated value of the corresponding measurement at the end of the driving cycle. During the first period

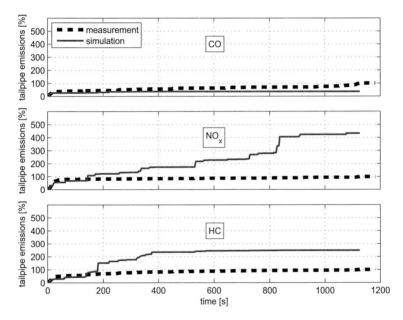

Figure 5.17.: Comparison of simulated and measured tailpipe emissions relative to the measured end amount during the NEDC [38]

of the urban part, the curve progressions of the cumulated pollutants show a good matching quality. Here, the monolith is passing its critical temperature range of the light-off. The good accordance belongs to the adjustments by the analysis of the light-off behavior in chapter 3.7. In the rest of the urban part, each pollutant shows different characteristics. The simulated tailpipe emissions of CO achieve the end value quickly and stay on a nearly constant level until the end of the driving cycle. The calculated conversion of HC is influenced by the decrease of H_2 and CO due to the ceria reactions of the oxygen storage module. Thus, more oxygen is remaining to oxidize C_3H_6 and C_3H_8. However, this indirect effect cannot prevent some noticeable breakthroughs at fuel-rich operation of the combustion engine. A stronger impact on the conversion by the oxygen storage model could be achieved by adding a cerium equation of direct HC oxidation. The cumulated NO_x emissions of the simulation show breakthroughs at every acceleration of the vehicle up to 50km/h. In this periods, increased loads of the combustion engine require a reduction of the residual gas concentration resulting in elevated raised engine-out emissions. The low conversion rate calculated for the component NO_x can be caused by the fact, that the oxygen storage module does not affect its reduction directly. Generally, a good matching quality for all pollutants can be achieved in the extra-urban part. Only in the simulated NO_x emissions, a significant breakthrough occurs at the first acceleration. Here, the same effects as already in the urban part, but with stronger impact are assumed ([38]).

In summary, there are still some deviations in the simulation results of the catalytic conversion during a driving cycle. Especially, in periods of high engine loads and fuel-rich gas compositions the conversion rate is underestimated and breakthroughs of HC and NO_x cannot be prevented. The transient behavior of the catalyst can be improved by adapting the oxygen storage model. One possibility is the expansion of the set of cerium equations. These reactions can be determined by additional laboratory scale experiments using synthesis gas and changing the air-fuel ratio according to Holder [63].

For further investigations, it is recommended to analyze the reactions directly affected by the ceria loading and storage capacity of the washcoat. Another aspect is the discretization of the grid points along the axial direction of the monolith. Analyses of the simulation model show that the main reactions of the mechanism only affect the first few millimeters at the entrance of the catalyst. Here, the storage level of oxygen has certainly more influence compared to the remaining area. Overall, a correct measuring of the oxygen storage capacity can enhance the quality of this module. Further improvements of the entire catalyst model can be achieved providing more detailed information about the monolith and its structure, and executing explicit measurements ([38]).

6

SUMMARY

In this thesis, a new coupling approach combining a calibration environment with a thermodynamic engine model and appending a virtual vehicle powertrain is implemented to simulate fuel consumption and in-cylinder gas emissions in various driving cycles. Hereby, the seamless coupling of different sub-models, as well as the extension of the predictive combustion model towards dynamic operation is the main challenge. In the context of fuel consumption and emissions analysis, this virtual approach offers the possibility to gain insights in the transient thermodynamic engine behavior in driving cycles. The simulation methodology enables the virtual calibration of an engine management system. This can be achieved by linking a functional model of an electronic control unit, a detailed combustion engine model including crank-angle based calculation of gas exchange and combustion and a mechanical vehicle powertrain model. In order to analyze in-cylinder emissions in driving cycles, different sub-models are implemented as subsequent process of the engine simulation.

For the purpose of this study, the ECU model includes all necessary functions to calculate the actuating variables of the combustion engine based on torque requirements of a virtual driver. Depending on the actual state of the engine (e.g. friction torque) and on the operating mode like catalyst heating, idle speed or fuel cutoff under overrun conditions, the torque module as central function is the connection between the driver input defined by the position of the gas pedal and other important parameters processed in the engine management like cylinder charging, ignition timing, and fuel injection. A hierarchical, modular structure is implemented to handle different calibration tasks and to enable the exchange/adaption of vehicle, powertrain, and engine. Furthermore, the model can be easily configured with an actual calibration version, adjusted to changes, and extended by functional expansions.

The thermodynamic processes are implemented by a 1D gas exchange simulation consisting of the entire intake/exhaust gas path of a turbo-charged DISI combustion engine modeled from air intake to tailpipe by physical objects. As a central issue of the presented novel methodology, the SI combustion process is calculated by an entrainment model in combination with a quasi-dimensional turbulence sub-model. A homogeneous fuel-mixture generation with a defined air-to-fuel ratio is assumed. Summarizing, the level of modeling represents a good compromise between computing time and quality of results. Due to the gas exchange simulation and the phenomenological combustion prediction, different calibration strategies of the system can be analyzed in various driving cycles. In order to investigate fuel consumption and in-cylinder emissions in testing cycles or real-life duty cycles, the powertrain of a virtual vehicle is integrated into the 1D simulation environment. Within the mechanical sub-model, the generated instantaneous torque of the combustion engine is converted via rotational powertrain elements to a longitudinal vehicle motion. For this purpose, the vehicle part is basically modeled rigid except of a flywheel to avoid the transformation of torque peaks from the thermodynamic engine simulation. Gearbox and differential are defined as kinematic transmission ratios with corresponding efficiency maps considering gear-dependent losses. Remaining drive-train resistances regarding acceleration, rolling, and aerodynamics are provided in the vehicle model as well.

Within the proposed simulation methodology, the ECU model is coupled to the detailed combustion engine/vehicle simulation directly via many actuator and sensor paths. Important is the time synchronization at the interfaces of the sub-models due to the fact that the ECU usually works in different grids in the range of milliseconds and the combustion engine simulation runs in crank train synchronized time steps. The coupled environment is extended by a driver's control to manage the gas and brake pedal dynamically according to the prescribed target velocity of the driving cycle.

The in-cylinder emissions are determined by additional sub-models implemented into the post-processing of the driving cycle simulations. The proposed novel model for the prediction of hydrocarbons emissions in turbocharged DISI engines includes the main formations effects piston crevice and oil layer. At cold engine start conditions, the additional source of wall quenching is considered as function of the engine temperature. The sub-model to predict the engine-out emissions of NO_x and CO is based on chemical reaction kinetics consisting of an iso-octane mechanism to determine the oxidation of fuel and a sub-mechanism for the calculation of the NO_x formation. Within a chemical solver, the conservation equations of mass, species, and energy are computed on the basis of an assumed two-zone approach. The conservation of pollutants in the TWC are described with the help of detailed chemical reaction equations, as well.

While the steady-state performance of the modeling approaches is matched during their calibration process, the separate verification of the transient results from each sub-model is demonstrated in detail. Furthermore, the interaction of the coupled environment including a virtual driver is validated successfully. Even the ECU functions are reduced by simplifications, the model produces results in high quality. The detailed engine model enables investigations of calibration effects to the gas exchange, but shows also some complications in the matching of the model parametrization. Nevertheless, satisfying simulation behavior is achieved in this work and presented in the validation. The vehicle dynamics are reproduced with high accuracy and in combination with the well-parametrized driver actuation, different speed profiles can be followed within the legal limits and the analysis of various driving cycles is enabled. In the context of the coupling approach, the path of potential error summation in the closed-loop system is identified and illustrated. Successful performance of the engine-out emissions prediction is demonstrated under transient conditions and sensitivities of each pollutant component can be reproduced by the sub-models.

Finally, the advantage of this complex linking technique is shown with the help of various application examples. The presented models can now be applied to upcoming changes in emission regulations regarding new driving cycles like RDE and WLTP. Beside further engine analysis, transient tailpipe emissions can be determined as well. The wide range of possible applications is proofed successfully in this work and approves the implementation of this coupled simulation methodology in future development challenges. For example, an extension of the in-cylinder emissions is scheduled by adding a virtual prediction of particle formation in driving cycles. Furthermore, simulation results can support the evaluation process of several concepts for engine actuation, calibration strategies, and exhaust gas aftertreatment.

BIBLIOGRAPHY

[1] Ahmed, S., Zeusch, T., Amneus, P., Blurock, E., Soyhan, H. and Mauss, F. (2002): *Validation of the iso-Octane/n-Heptane mechanism PL 2.0 in HCCI and shock tube calculations*. Planet D4 report, The European Community.

[2] Angelidis, T.N. and Sklavounos, S.A. (1995): *A SEM-EDS study of new and used automotive catalysts*. Applied Catalysis A: General, vol. 133(1), pp. 121–132.

[3] Auer, M. (2010): *Erstellung phänomenologischer Modelle zur Vorausberechnung des Brennverlaufes von Magerkonzept-Gasmotoren*. Ph.D. thesis, TU München, Munic, Germany.

[4] Bartholomew, C. H. (2001): *Mechanisms of catalyst deactivation*. Applied Catalysis A: General, vol. 212(1), pp. 17–60.

[5] Bielaczyc, P., Woodburn, J. and Szczotka, A. (2014): *The WLTP as a new tool for the evaluation of CO2 emissions*. FISITA 2014 World Automotive Congress, Maastricht, Netherlands, (F2014-CET-139).

[6] Bird, R. B., Stewart, W. E. and Lightfoot, E. N. (2006): *Transport Phenomena, Revised 2nd Edition*. John Wiley & Sons, Inc., Hoboken, NJ, USA.

[7] Blizard, N. C. and Keck, J. C. (1974): *Experimental and theoretical investigation of turbulent burning model for internal combustion engines*. SAE Technical Paper, (740191).

[8] BMW Group (2014): *Technische Daten. MINI John Cooper Works, MINI John Cooper Works Automatik*. MINI Medieninformation 12/2014.

[9] Borrmeister, J. and Hübner, W. (1997): *Einfluss der Brennraumform auf HC-Emission und den Verbrennungsablauf*. MTZ - Motortechnische Zeitschrift, vol. 58(7/8), pp. 408–414.

[10] Brandt, M., Raatz, T., Lejsek, D. and Betzler, P. (2010): *Chances and Limitations of Scavenging for extreme downsized Engines*. 15. Aufladetechnische Konferenz, Dresden, Germany.

[11] Chatterjee, D. (2001): *Detaillierte Modellierung von Abgaskatalysatoren*. Ph.D. thesis, Ruprecht-Karls-Universität Heidelberg, Heidelberg, Germany.

[12] Cheng, W. K., Hamrin, D., Heywood, J. B., Hochgreb, S., Min, K. and Norris, M. (1993): *An Overview of Hydrocarbon Emissions Mechanisms in Spark-Ignition Engines*. SAE Technical Paper, (932708).

[13] Commitee to Assess Fuel Economy Technologies for Medium- and Heavy-Duty Vehicles (2010): *Technologies and Approaches to Reducing the Fuel Consumption of Medium- and Heavy-Duty Vehicles*. The National Academic Press, Washington, D.C., USA.

[14] Curran, H. J., Gaffuri, P., Pitz, W. J. and Westbrook, C. K. (2002): *A Comprehensive Modeling Study of iso-Octane Oxidation.* Combustion and Flame, vol. 129, pp. 253–280.

[15] Darnton, N. J. (1997): *Fuel Consumption and Pollutant Emissions of Spark Ignition Engines During Cold-Started Drive Cycles.* Ph.D. thesis, University of Nottingham, Nottingham, UK.

[16] Dent, J. C. and Lakshminarayanan, P. A. (1983): *A Model for Adsorption and Desorption of Fuel Vapour by Cylinder Lubricating Oil Films and its Contribution to Hydrocarbon Emissions.* SAE Technical Paper, (830652).

[17] D'Errico, G., Ferrari, G., Onorati, A. and Cerri, T. (2002): *Modeling the Pollutant Emissions from a S.I. Engine.* SAE Technical Paper, (2002-01-0006).

[18] D'Errico, G. and Onorati, A. (2006): *Thermo-Fluid Dynamic Modelling of a Six-Cylinder Spark Ignition Engine with a Secondary Air Injection System.* International Journal of Engine Research, vol. 7(1), pp. 1–16, ISSN 1468-0874.

[19] Deutschmann, O. (2001): *Interactions between transport and chemistry in catalytic reactors.* Habilitationsschrift, Ruprecht-Karls-Universität Heidelberg, Heidelberg, Germany.

[20] Deutschmann, O., Nieken, U. and Eigenberger, G. (2010): *Korrelation und Modellierung des Katalysatorumsatzverhaltens bei Variation der Edelmetallbeladung, des OSC und des Alterungszustandes.* Abschlussbericht FVV-Vorhaben 953.

[21] Doll, M. (2010): *Skript: Motorsteuerung.* Fachhochschule München, Munic, Germany.

[22] Dorsch, M. (2011): *Echtzeitfähige 1D Thermodynamikmodellierung für die Motorapplikation direkteinspritzender Benzinmotoren.* Diplomarbeit, Universität Stuttgart, Stuttgart, Germany.

[23] Dorsch, M., Matros, K., Neumann, J. and Hasse, C. (2013): *Simulation der Dynamik in Fahrzeugantrieben mit aufgeladenen Ottomotoren – Methodik und Analyse.* 9. Tagung Dynamisches Gesamtsystemverhalten von Fahrzeugantrieben, Starnberg, Germany.

[24] Dorsch, M., Neumann, J. and Hasse, C. (2014): *Detailed modeling of SI engines in driving cycle simulations for fuel consumption analysis.* FISITA 2014 World Automotive Congress, Maastricht, Netherlands, F2014-CET-017.

[25] Dorsch, M., Neumann, J. and Hasse, C. (2014): *Nutzung der Ladungswechsel- und Motorprozesssimulation zur Gesamtsystembewertung von CO2- und Rohemissionen in Fahrzyklen.* 7. MTZ-Fachtagung Ladungswechsel im Verbrennungsmotor, Stuttgart, Germany.

[26] Dorsch, M., Neumann, J. and Hasse, C. (2015): *Application of a Phenomenological Model for the Engine-out Emissions of Unburned Hydrocarbons in Driving Cycles.* Journal of Energy Resources Technology, vol. 138(2), pp. 022201–10.

[27] Drangel, H., Nordin, H., Johansson, P. and Koenigstein, A. (2007): *Charging System for a High Performance SI-Engine - Technology and Methods.* 12. Aufladetechnische Konferenz, Dresden, Germany.

[28] Economic Commission for Europe (2014): *Proposal for a new global technical regulation on the Worldwide harmonized Light vehicles Test Procedure (WLTP)*. World Forum for Harmonization of Vehicle Regulations 162nd session, Geneva, Switzerland.

[29] Eichlseder, H., Klüting, M. and Piock, W. (2008): *Grundlagen und Technologien des Ottomotors - Der Fahrzeugantrieb*. Springer-Verlag, Berlin, Germany.

[30] El-Mahallawy, F. and Habik, S. E-Din (2002): *Fundamentals and Technology of Combustion*. Elsevier, Amsterdam, Netherlands.

[31] Ertl, G., Künzinger, H. and Weitkamp, J. (1999): *Environmental Catalysis*. Wiley-VCH Verlag GmbH, Weinheim, Germany.

[32] ETAS (Abgerufen am: 16.05.2012): *Applikation von elektronischen Fahrzeugsystemen*. ETAS GmbH, http://www.etas.com.

[33] Europäisches Parlament (1991): *Richtlinie 91/441/EWG des Rates vom 26. Juni 1991 zur Änderung der Richtlinie 70/220/EWG zur Angleichung der Rechtsvorschriften der Mitgliedstaaten ber Manahmen gegen die Verunreinigung der Luft durch Emissionen von Kraftfahrzeugen*. Amtsblatt der Europäischen Union, (L242), pp. 1–106.

[34] Fenimore, C. P. (1971): *Formation of Nitric Oxide in Premixed Hydrocarbon Flames*. Symposium (International) on Combustion, vol. 13(1), pp. 373–380, ISSN 00820784.

[35] Friedrich, I., Buchwald, R., Stlting, E. and Sommer, A. (2009): *Das virtuelle Fahrzeug - Transiente Simulation für den dieselmotorischen Entwicklungsprozess*. MTZ - Motortechnische Zeitschrift, vol. 70(12), pp. 922–929.

[36] Fritsche, C. (2010): *Ein Beitrag zur prädiktiven Regelung verbrennungsmotorischer Prozesse*. Ph.D. thesis, Universität Rostock, Rostock, Germany.

[37] Frölund, K. and Schramm, J. (1997): *Simulation HC-Emissions from SI-Engines - A Parametric Study*. SAE Technical Paper, (972893).

[38] Frommater, S. (2013): *Simulation Of The Three-Way Catalyst Conversion Behavior On Engine Test Benchs And In The European Driving Cycle*. Master thesis, Technische Universität Bergakademie Freiberg, Freiberg, Germany.

[39] Froschhammer, F., Rabenstein, F. and Mathiak, D. (2006): *Hochdynamische Prüfstände - Ein Werkzeug für die Instationärapplikation*. ATZ/MTZ Konferenz - Motor, Motorenentwicklung auf dynamischen Prüfständen, Wiesbaden, Germany.

[40] Gamma Technologies Inc. (2011): *GT-Power User´s Manual*. GT-Suite version 7.2.

[41] Gatellier, B., Trapy, J., Herrier, D., Quelin, J. M. and Galliot, F. (1992): *Hydrobarbon Emissions of SI Engines as Influenced by Fuel Absorption-Desorption in Oil Films*. SAE Technical Paper, (920095).

[42] Gerhardt, J., Benninger, N. and Hess, W. (1997): *Drehmomentbasierte Funktionsstruktur der elektrischen Motorsteuerung als neue Basis für Triebstrangsysteme*. 6. Aachener Kolloquium, Fahrzeug und Motorentechnik, Aachen, Germany.

[43] Gerhardt, J., Hönninger, H. and Bischof, H. (1998): *A new approach to fufunction and software structure for engine management systems - Bosch ME7.* SAE Congress & Exposition, (98P-178 (49)).

[44] Golloch, R. (2005): *Downsizing bei Verbrennungsmotoren. Ein wirkungsvolles Konzept zur Kraftstoffverbrauchssenkung.* Springer-Verlag, Berlin, Germany.

[45] Golovitchev, V. (abgerufen: Juni 2015): *Iso-octane mechanism.* www.tfd.chalmers.se/~valeri/MECH.html.

[46] Göschel, B. (2006): *Skript: Die Zukunft des ottomotorischen Antriebs.* Technische Universität Graz, Graz, Austria.

[47] Gotter, A. and Pischinger, S. (2006): *Motorsteuerung mit Eingriffsmöglichkeit über ein Applikationssystem für stationär betriebene DI-Benzin-Motoren.* Abschlussbericht FVV-Vorhaben 843.

[48] Grasreiner, S. (2012): *Combustion modeling for virtual SI engine calibration with the help of 0D/3D methods.* Ph.D. thesis, Technische Universität Bergakademie Freiberg, Freiberg, Germany.

[49] Grasreiner, S., Neumann, J., Luttermann, C., Wensing, M. and Hasse, C. (2014): *A quasi-dimensional model of turbulence and global charge motion for SI engines with fully-variable valve-trains.* Int. J. Engine Res., vol. 15(7), pp. 805–816.

[50] Grasreiner, S., Neumann, J., Wensing, M. and Hasse, C. (2015): *A quasi-dimensional model of the ignition delay for combustion modeling in SI engines.* Journal Engine Gas Turbines Power, vol. 137(7)(071502-1).

[51] Grill, M. (2006): *Objektorientierte Prozessrechnung von Verbrennungsmotoren.* Ph.D. thesis, Universität Stuttgart, Stuttgart, Germany.

[52] Grill, M., Billinger, T. and Bargende, M. (2006): *Quasi-Dimensional Modeling of Spark Ignition Engine Combustion with Variable Valve Train.* SAE Technical Paper, (2006-01-1107).

[53] Hagelüken, C. (2010): *Autoabgaskatalysatoren: Grundlagen, Herstellung, Entwicklung, Recycling, Ökologie.* expert verlag GmbH, Renningen, Germany.

[54] Hasewend, W. (2001): *AVL Cruise – Driving performance and fuel consumption simulation.* ATZ worldwide, vol. 103(5), pp. 10–13.

[55] Hasse, C., Bollig, M., Peters, N. and Dwyer, H. A. (2000): *Quenching of Laminar Iso-Octane Flames at Cold Walls.* Combustion and Flame, vol. 122, pp. 117–129.

[56] Haupt, C. (2013): *Ein multiphysikalisches Simulationsmodell zur Bewertung von Antriebs- und Wärmemanagementkonzepten im Kraftfahrzeug.* Ph.D. thesis, TU München, Munic, Germany.

[57] Hayes, R. E. and Kolaczkowski, S. T. (1998): *Introduction to Catalytic Combustion.* CRC Press, Danvers, MA, USA.

[58] Herz, R. K. (1981): *Dynamic behavior of automotive catalysts. 1. Catalyst oxidation and reduction.* Industrial & Engineering Chemistry Product Research and Development, vol. 20(3), pp. 451–457.

[59] Heywood, J. B. (1988): *Internal Combustion Engine Fundamentals.* McGraw-Hill Inc., New York City, NY, USA.

[60] Hiereth, H. and Prenninger, P. (2003): *Aufladung der Verbrennungskraftmaschine.* Springer Verlag, Wien, Austria.

[61] Hochgreb, S. (1998): *Combustion-Related Emissions in SI Engines*, chap. Handbook of Air Pollution from Internal Combustion Engines: Pollutant Formation and Control. Eran Sher, Academic, Boston, USA.

[62] Hoebink, J. H. B. J. and Harmsen, J. M. A. (2006): *Modeling of automotive exhaust gas converters.* Structured Catalysts and Reactors 2, pp. 311–354.

[63] Holder, R. (2008): *A Global Reaction Mechanism for Transient Simulations of Three-Way Catalytic Converters.* Ph.D. thesis, RWTH Aachen, Aachen, Germany.

[64] Holder, R., Bollig, M., Anderson, D.R. and Hochmuth, J.K. (2006): *A discussion on transport phenomena and three-way kinetics of monolithic converters.* Chemical Engineering Science, vol. 61(24), pp. 8010–8027.

[65] Huang, C., Golovitchev, V. and Lipatnikov, A. (2010): *Chemical Model of Gasoline-Ethanol Blends for Internal Combustion Engine Applications.* SAE Technical Paper, (2010-01-0543).

[66] Huang, Z., Pan, K., Li, J., Zhou, L. and Jiang, D. (1996): *An Investigation on Simulation Models and Reduction Methods of Unburned Hydrocarbon Emissions in Spark Ignition Engines.* Combustion Science and Technology, vol. 115(1-3), pp. 105–123, ISSN 0010-2202.

[67] Ignatov, D., Küpper, C., Pischinger, S., Bahn, M., Betton, W., Rütten, O. and Weinowski, R. (2010): *Catalyst Aging Method for Future Emissions Standard Requirements.* SAE Technical Paper, (2010-01-1272).

[68] Isermann, R. (ed.) (2003): *Modellgestützte Steuerung, Regelung und Diagnose von Verbrennungsmotoren.* Springer Verlag, Berlin, Germany.

[69] Isermann, R. (ed.) (2010): *Elektronisches Management motorischer Fahrzeugantriebe.* 1. Auflage, Vieweg+Teubner, Wiesbaden, Germany.

[70] Janssen, C. (2010): *Möglichkeiten zur Prädiktion von unverbrannten Kohlenwasserstoffen in einem direkteinspritzenden Ottomotor.* Ph.D. thesis, Universität Rostock, Rostock, Germany.

[71] Jippa, K.-N. (2003): *Onlinefhige, thermodynamikbasierte Ansätze für die Auswertung von Zylinderdruckverläufen.* Ph.D. thesis, Universität Stuttgart.

[72] Jobson, E., Laurell, M., Högberg, E. and Bernler, H. et al. (1993): *Deterioration of Three-Way Automotive Catalysts, Part I - Steady State and Transient Emission of Aged Catalyst.* SAE Technical Paper, (930937).

[73] Joos, F. (2006): *Technische Verbrennung, Verbrennungstechnik, Verbrennungsmodellierung, Emissionen.* Springer Verlag, Berlin, Germany.

[74] Kang, S. B., Han, S. J., Nam, S. B., Nam, In-Sik, Cho, B. K., Kim, C. H. and Oh, S. H. (2012): *Activity function describing the effect of Pd loading on the catalytic performance of modern commercial TWC.* Chemical Engineering Journal, vol. 207-208, pp. 117–121.

[75] Kau, H.-P. (2010): *Skript: Grundlagen der Strömungsmaschinen.* Technische Universität München.

[76] Keil, F. J. (1999): *Diffusion and reaction in porous networks.* Catalysis Today, vol. 53(2), pp. 245–258.

[77] Kfz-Technik (Abgerufen am: 07.05.2012): *Valvetronic (BMW).* Kfz-Technik Wiesinger, http://www.kfztech.de/kfztechnik/motor/steuerung/valvetronic.htm.

[78] Kirsten, K. (2011): *Variabler Ventiltrieb im Spannungsfeld von Downsizing und Hybridantrieb.* 32. Internationales Wiener Motorensymposium, Vienna, Austria.

[79] Klauer, N., Klüting, M., Schünemann, E., Schwarz, C. and Steinparzer, F. (2013): *BMW TwinPower Turbo Gasoline Engine Technology - Enabling Compliance with worldwide Exhaust Gas Emissions Requirements.* 34. Internationales Wiener Motorensymposium, Vienna, Austria.

[80] Klauer, N., Klüting, M., Steinparzer, F. and Kretschmer, J. (2009): *Innovative Aufladung, Variable Ventiltriebe und 8-Gang Getriebe - eine neue Generation von Antrieben.* Aachener Kolloquium Fahrzeug- und Motorentechnik, Aachen, Germany.

[81] Klauer, N., Klüting, M., Steinparzer, F. and Unger, H. (2009): *Aufladung und variable Ventiltriebe – Verbrauchstechnologien für den weltweiten Einsatz.* 30. Internationales Wiener Motorensymposium, Vienna, Austria.

[82] Klingenberg, H. (2013): *Automobil-Metechnik: Band C: Abgasmetechnik.* Springer Verlag, Berlin, Germany.

[83] Knoll, U., Zaglauer, S., Grasreiner, S. and Hasse, C. (2011): *Automatische Kalibrierung von Simulationsmodellen zur virtuellen Motorapplikation.* AUTOREG - Steuerung und Regelung von Fahrzeugen und Motoren, Baden-Baden, Germany.

[84] Koltsakis, G. C., Konstantinidis, P. A. and Stamatelos, A. M. (1997): *Development and application range of mathematical models for 3-way catalytic converters.* Applied Catalysis B: Environmental, vol. 12(2), pp. 161–191.

[85] Koltsakis, G. C. and Stamatelos, A. M. (1997): *Catalytic automotive exhaust aftertreatment.* Progress in Energy and Combustion Science, vol. 23(1), pp. 1–39.

[86] Koop, J. (2008): *Detaillierte Modellierung der Pt-katalysierten Schadstoffminderung in Abgasen moderner Verbrennungsmotoren.* Ph.D. thesis, Universität Fridericiana Karlsruhe (TH), Karlsruhe, Germany.

[87] Kratzsch, M., M.Günther, Elsner, N. and Zwahr, S. (2009): *Modellansätze für die virtuelle Applikation von Motorsteuergeräten.* MTZ - Motortechnische Zeitschrift, vol. 70(9), pp. 664–670.

[88] Kreith, F. (1960): *Principles of Heat Transfer*. International Textbook Co, Scranton, PA, USA.

[89] Kruse, T., Ulmer, H. and Lange, T. (2011): *Einsatz neuer statischer Lernverfahren in der Applikation von Motorsteuerungen*. VDI-Berichte Nr. 2135, S. 655ff.

[90] Kwon, H. J., Hyun Baik, J., Tak Kwon, Y., Nam, I.-S. and Oh, S. H. (2007): *Detailed reaction kinetics over commercial three-way catalysts*. Chemical Engineering Science, vol. 62(18-20), pp. 5042–5047.

[91] Lanzerath, P. (2012): *Alterungsmechanismen von Abgaskatalysatoren für Nutzfahrzeug-Dieselmotoren*. Ph.D. thesis, Technische Universität Darmstadt, Darmstadt, Germany.

[92] Lassi, U (2003): *Deactivation correlations of Pd/Rh three-way catalysts designed for Euro IV emission limits: Effect of ageing atmosphere, temperature and time*. Ph.D. thesis, University of Oulu, Oulu, Finland.

[93] Lavoie, G. A. (1978): *Correlations of Combustion Data for S.I. Engine Calculations – Laminar Flame Speed, Quench Distance and Global Reaction Rates*. SAE Technical Paper, (780229).

[94] Lavoie, G. A., Heywood, J. B. and Keck, J. C. (1970): *Experimental and Theoretical Investigation of Nitric Oxide Formation in Internal Combustion Engines*. Combustion Science and Technology, vol. 1, pp. 313–326.

[95] Lichtenthäler, D. (2001): *Prozessmodelle mit integrierten neuronalen Netzen zur Echtzeitsimulation und Diagnose von Verbrennungsmotoren*. Fortschrittsberichte VDI Reihe 12 Nr. 454, VDI Verlag, Düsseldorf, Germany.

[96] Liesen, W. (1997): *Einfluss externer und interner Abgasrückführung auf das HC Emissionsprofil im ottomotorischen Abgas*. Ph.D. thesis, RWTH Aachen, Aachen, Germany.

[97] Linna, J. R. (1997): *Contribution of Oil Layer Mechanism to the Hydrocarbon Emissions from Spark-Ignition Engines*. Ph.D. thesis, Massachusetts Institute of Technology, Cambridge, MA, USA.

[98] Linna, J.-R., Malberg, H., Bennett, P. J., Palmer, J., Tian, T. and Cheng, W. K. (1997): *Contribution of Oil Layer Mechanism to the Hydrocarbon Emissions from Spark-Ignition Engines*. SAE Technical Paper, (972892).

[99] Linse, D. (2013): *Modeling and Simulation of Knock and Emissions in Turbocharged Direct Injection Spark Ignition Engines*. Ph.D. thesis, TU Bergakademie Freiberg, Freiberg, Germany.

[100] Luttermann, C., Missy, S., Schwarz, C. and Klauer, N. (2006): *High Precision Injection in Verbindung mit Aufladung am Beispiel des neuen BMW TwinTurbo Ottomotors*. Aachener Kolloquium Fahrzeug- und Motorentechnik, Aachen, Germany.

[101] Marotta, A., Pavlovic, J., Ciuffo, B., Serra, S. and Fontaras, G. (2015): *Gaseous Emissions from Light-Duty Vehicles: Moving from NEDC to the New WLTP Test Procedure*. Environmental Science and Technology, vol. 49(14), pp. 8315–8322.

[102] Martin, L., Arranz, J.L., Prieto, O., Trujillano, R., Holgado, M.J., Galán, M.A. and Rives, V. (2003): *Simulation three-way catalyst ageing*. Applied Catalysis B: Environmental, vol. 44(1), pp. 41–52.

[103] MathWorks (Abgerufen: 23.05.2012): *Simulink*. The MathWorks Inc., http://www.mathworks.de/products/simulink/.

[104] Maurer, F. (2013): *Untersuchung des Spülverhaltens eines Motors für die eindimensionale Ladungswechselsimulation*. Diplomarbeit, Hochschule für angewandte Wissenschaften FH München, Munic, Germany.

[105] Meder, G., Mitterer, A.and Konrad, H., Krämer, G. and Siegl, N. (2006): *Entwicklung und Applikation von modellbasierten Steuergerätefunktionen am Beispiel der neuen BMW Reihensechszylindermotoren mit Valvetronic*. 3. Fachtagung Steuerung und Regelung von Fahrzeugen und Motoren - AUTOREG, Wiesloch, Germany, ISSN 0178-2312.

[106] Meisberger, D., Albert, C. and Bourdon, K. (1997): *Die neue Motorsteuerung ME7.2 von Bosch für den neuen BMW V8-Motor*. MTZ - Motortechnische Zeitschrift, vol. 58(12), pp. 826–834.

[107] Mencher, B., Jessen, H., Kaiser, L. and Gerhardt, J. (2001): *Preparing for CARTRONIC - Interface and New Strategies for Torque Coordination and Conversion in a Spark Ignition Engine-Management System*. SAE Technical Paper, (2001-01-0268).

[108] Merker, G. P. and Schwarz, C. (2009): *Grundlagen Verbrennungsmotoren: Simulation der Gemischbildung, Verbrennung, Schadstoffbildung und Aufladung*. Vieweg+Teubner, Wiesbaden, Germany.

[109] Merker, G. P., Schwarz, C., Stiesch, G. and Otto, F. (2006): *Verbrennungsmotoren: Simulation der Verbrennung und Schadstoffbildung*. 3. überarb. und akt. Aufl., Vieweg+Teubner Verlag, Wiesbaden, Germany.

[110] Metghalchi, M. and Keck, J. C. (1982): *Burning velocities of mixtures of air and methanol, isooctane and indolene an high pressure and temperature*. Combustion and Flame, vol. Band 48, pp. 191–210.

[111] Miklautschitsch, M. (2011): *Niedrigstemissionskonzept auf Basis eines abgasturboaufgeladenen Ottomotors mit Direkteinspritzung, vollvariablem Ventiltrieb und Sekundärlufteinblasung*. Ph.D. thesis, Karlsruher Institutes für Technologie (KIT), Karlsruhe, Germany.

[112] Millo, F., Rolando, L. and Andreata, M. (2011): *Numerical Simulation for Vehicle Powertrain Development*, numerical analysis - theory and application 24, pp. 519–540. ISBN: 978-953-307- 389-7, InTech.

[113] Min, K. (1994): *The Effects of Crevices on the Engine-Out Hydrocarbon Emissions in Spark Ignition Engines*. Ph.D. thesis, Massachusetts Institute of Technology, Cambridge, MA, USA.

[114] Min, K. and Sloan, W. K. C. (1995): *Oxidation of the Piston Crevice Hydrocarbon During the Expansion Process in a Spark Ignition Engine*. vol. 106(617), pp. 307–326.

[115] Mladenov, N. (2009): *Modellierung von Autoabgaskatalysatoren*. Ph.D. thesis, Universität Karlsruhe (TH), Karlsruhe, Germany.

[116] Mock, P., Kühlwein, J., Tietge, U., Franco, V., Bandivadekar, A. and German, J. (2014): *The WLTP: How a new test procedure for cars will affect fuel consumption values in the EU*. ICCT, (WORKING PAPER 2014-9).

[117] Morel, T., Rackmil, C. I., Keribar, R. and Jennings, M. J. (1988): *Model for Heat Transfer and Combustion in Spark Ignited Engines and its Comparsion with Experiments.*. SAE Technical Paper, 880198.

[118] Nefischer, A. (2009): *Quasidimensionale Modellierung turbulenzgetriebener Phänomene in Ottomotoren*. Ph.D. thesis, Technische Universität Graz, Graz, Austria.

[119] Nefischer, A., Neumann, J., Stanciu, A. and Wimmer, A. (2010): *Comparison and Application of different phenomenological Combustion-Models for turbocharged SI-Engines*. FISITA World Automotive Congress, Budapest, Hungary.

[120] Nelles, O. (2001): *Nonlinear System Identification*. Springer-Verlag, Berlin, Germany.

[121] Neumeister, J., Taylor, J. and Gurney, D. (2007): *Virtual Air Path Calibration of a Multi Cylinder High Performance GDI Engine using 1D Cycle Simulation*. SAE Technical Paper, (2007-01-0490).

[122] Nietschke, W., Schultalbers, M. and Magnor, O. (2005): *Der zunehmende Einfluss der Simulation auf die Entwicklung der Motorsteuerung*. 26. Internationales Wiener Motorsymposium, Vienna, Austria.

[123] Nijhuis, T. Alexander, Beers, Annemarie E. W., Vergunst, Theo, Hoek, Ingrid, Kapteijn, Freek and Moulijn, Jacob A. (2001): *Preparation of monolithic catalysts*. Catalysis Reviews, vol. 43(4), pp. 345–380.

[124] Nijis, M., Sternberg, P., Wittler, M. and Pischinger, S. (2010): *Steuergerätefähige Luftpfadmodelle für Ottomotoren mit erweiterter Ventiltriebsvariabilität*. MTZ - Motortechnische Zeitschrift, vol. 71(11), pp. 824–831.

[125] Norris, M. G. and Hochgreb, S. (1994): *Novel Experiment on In-Cylinder Desorption of Fuel from Oil Layer*. SAE Technical Paper, (941963).

[126] Norris, M. G. and Hochgreb, S. (1996): *Extent of Oxidation from the Lubricant Oil Layer in Spark-Ignition Engines*. SAE Technical Paper, (960069).

[127] Oliveira, I. B. and Hochgreb, S. (1999): *Effect of Operating Conditions and Fuel Type on Crevice HC Emissions: Model Results and Comparison with Experiments*. SAE Technical Paper, (1999-01-3578).

[128] Pacejka, H. (1996): *The Tyre as a Vehicle Component*. 26. FISITA Congress, Prague, Czech Republic.

[129] Pester, W. (2014): *Neuer Welt-Sprit-Test liefert 20 Prozent höhere Werte*. Global Press Nachrichtenagentur und Informationsdienste GmbH, Düsseldorf, Germany.

[130] Peters, N. (2000): *Turbulent Combustion*. Wiley, New York, USA.

[131] Petzold, L. R. (1982): *A Description of DASSL: A Differential/Algebraic System Solver*. Tech. Rep. SAND82-8637, Sandia National Laboratories.

[132] Pontikakis, G. and Stamatelos, A. (2001): *Mathematical modelling of catalytic exhaust systems for EURO-3 and EURO-4 emissions standards*. Proceedings of the Institution of Mechanical Engineers, Part D: Journal of Automobile Engineering, vol. 215(9), pp. 1005–1015.

[133] Ramanathan, K. and Sharma, C. S. (2011): *Kinetic Parameters Estimation for Three Way Catalyst Modeling*. Industrial & Engineering Chemistry Research, vol. 50(17), pp. 9960–9979.

[134] Reulein, C. and Schwarz, C. (2009): *Senkung der CO2-Emissionen Durch Ladungswechsel*. MTZ - Motortechnische Zeitschrift, vol. 71(11), pp. 760–765.

[135] Ricardo Inc. and Systems Research and Applications Corporation (SRA) (2011): *Computer Simulation of Light-Duty Vehicle Technologies for Greenhouse Gas Emission Reduction in the 2020-2025 Timeframe*. EPA-420-R-11-020.

[136] Rieckmann, C. and Keil, F. J. (1999): *Simulation and experiment of multicomponent diffusion and reaction in three-dimensional networks*. Chemical Engineering Science, vol. 54(15), pp. 3485–3493.

[137] Riegler, U. G. (1999): *Berechnung der Verbrennung und der Schadstoffbildung in Ottomotoren unter Verwendung detaillierter Reaktionsmechanismen*. Ph.D. thesis, Universität Stuttgart, Stuttgart, Germany.

[138] Robert Bosch GmbH (ed.) (1998): *Ottomotor-Management*. 1. Auflage, Friedrich Vieweg & Sohn Verlagsgesellschaft mbH, Wiesbaden, Germany.

[139] Robert Bosch GmbH (ed.) (2005): *Ottomotor-Management - Systeme und Komponenten*. 3. Auflage, Friedrich Vieweg & Sohn Verlagsgesellschaft mbH, Wiesbaden, Germany.

[140] Roithmeier, C. (2011): *Virtuelle Applikation von Motorsteuerungsfunktionen am Beispiel der Lasterfassung und der Fahrdynamikfunktionen*. Ph.D. thesis, Karlsruher Institut für Technologie (KIT), Karlsruhe, Germany.

[141] Sailer, T., Bucher, S., Durst, B. and Schwarz, C. (2011): *Simulation des Verdichterverhaltens von Abgasturboladern*. MTZ - Motortechnische Zeitschrift, vol. 72(01), pp. 28–33.

[142] Sailer, T., Hausner, O., Reulein, C. and Schwarz, C. (2013): *Neue Methoden in der Turboladersimulation für die Auslegung der BMW TwinPower Turbomotoren*. 6. MTZ-Fachtagung Ladungswechsel in Verbrennungsmotoren, Stuttgart, Germany.

[143] Sams, T. (Wintersemester 2010/2011): *Vorlesung: Schadstoffbildung und Emissionierung bei Kfz*. TU Graz, Graz, Austria.

[144] Santavicca, D. A., Liou, D. and North, G. L. (1990): *A Fractal Model of Turbulent Flame Kernel Growth*. SAE Technical Paper, (900024).

[145] Schade, H., Kunz, E., Kameier, F. and Paschereit, C. O. (2013): *Strömungslehre*. Walter de Gruyter GmbH, Berlin, Germany.

[146] Schmid, A. and Bargende, M. (2011): *Efficiency Optimization of SI-Engines in Real World Driving Conditions*. 20th Aachen Colloquium Automobile and Engine Technology, Aachen, Germany.

[147] Schmid, A., Gril, M., Berner, H.-J. and Bargende, M. (2009): *Transient and Map-Based Driving Cycle Calculation with GT-SUITE*. GT-User Conference, Frankfurt am Main, Germany.

[148] Schmid, A., Grill, M., Berner, H.-J. and Bargende, M. (2010): *Transiente Simulation mit Scavenging beim Turbo-Ottomotor*. MTZ - Motortechnische Zeitschrift, vol. 71(11), pp. 766–772.

[149] Schmid, A., Grill, M., Berner, H.-J. and Bargende, M. (2011): *Virtuelle Optimierung an einem Twin-Turbo Dreizylinder-Ottomotor im FTP75*. 3. Tagung: Motorprozessrechnung und Aufladung, Berlin, Germany.

[150] Schramm, J. and Sorenson, S. C. (1990): *A Model for Hydrocarbon Emissions from SI Engines*. SAE Technical Paper, (902169).

[151] Schreiner, J. (2015): *Deutsche OEM hinken CO2-Zielen hinterher*. Automobil Industrie, Würzburg, Germany.

[152] Seider, G. and Bet, F. (2010): *Neue Simulationstechniken - Potenziale für den virtuellen Produktentstehungsprozess*. Haus der Technik - 7. Tagung - Wärmemanagement des Kraftfahrzeugs, Berlin, Germany.

[153] Shyy, W. and Jr., T. C. Adamson (1983): *Analysis of Hydrocarbon Emissions From Conventional Spark-Ignition Engines*. Combustion Science and Technology, vol. 33, pp. 245–260.

[154] Sigg, D., Schneider, J. and Andrieux, G. (2010): *Neuer Valvetronic-Aktuator für den Turbomotor des Mini*. MTZ - Motortechnische Zeitschrift, vol. 71(10), pp. 712–717.

[155] Sodre, J. R. (1998): *A Parametric Model for Spark Ignition Engine Turbulent Flame Speed*. SAE Technical Paper, (982920).

[156] Sodre, J. R. (1999): *Further Improvements on a HC Emissions Model: Partial Burn Effects*. SAE Technical Paper, (1999-01-0222).

[157] Sodre, J. R. and Yates, D. A. (1997): *An Improved Model for Spark Ignition Engine Exhaust Hydrocarbon*. SAE Technical Paper, (971011).

[158] Spicher, U., Feng, B. and Kölmel, A. (1999): *HC-Rohemissionen beim Kaltstart in der Warmlaufphase sowie bei Last- und Drehzahlsprüngen*. Forschungsbericht FZKA-BWPLUS, Förderkennzeichen: PEF 3 96 003, Institut für Kolbenmaschinen, Universität Karlsruhe.

[159] Steinparzer, F., Schwarz, C., Brüner, T. and Mattes, W. (2014): *The new BMW 3- and 4-Cylinder Petrol Engines with Twin Power Turbo Technology*. 35. Internationales Wiener Motorensymposium, Vienna, Austria.

[160] Suck, G. (2001): *Untersuchung der HC-Quellen an einem Ottomotor mit Direktein-spritzung.* Ph.D. thesis, Otto-von-Guericke-Universität Magdeburg, Magdeburg, Germany.

[161] Tabaczynski, R. J., Ferguson, C. R. and Radhakrishnan, K. (1977): *A turbulent entrainment model for spark-ignition engine combustion.* SAE Technical Paper, (770647).

[162] Tallio, K. V. and Colella, P. (1997): *A Multi-Fluid CFD Turbulent Entrainment Combustion Model: Formulation and One-Dimensional Results.* SAE Technical Paper, 972880.

[163] Tennekes, H. and Lumley, J. L. (1972): *A First Course in Turbulence.* MIT Press, Cambridge, MA, USA.

[164] Trinker, F. H., Chen, J. and Davis, G. C. (1993): *A Feedgas HC Emission Model for SI Engines Including Partial Burn Effects.* SAE Technical Paper, (932705).

[165] Tsinoglou, D. N., Koltsakis, G. C. and Peyton Jones, J. C. (2002): *Oxygen storage modeling in three-way catalytic converters.* Industrial & Engineering Chemistry Research, vol. 41(5), pp. 1152–1165.

[166] Tsinoglou, D. N. and Weilenmann, M. (2009): *A simplified three-way catalyst model for transient hot-mode driving cycles.* Industrial & Engineering Chemistry Research, pp. 1772–1785.

[167] Ullmann, M. (2012): *Development of a Multi-Physics Tool for the Simulation of Chemical and Physical Processes in Three-Way Catalysts.* Master thesis, TU Bergakademie Freiberg, Freiberg, Germany.

[168] UNECE Global Technical Regulation No. 15 (2015): *Worldwide Harmonized Light Vehicles Test Procedure.* UNECE.

[169] Unger, H., Schwarz, C., Schneider, J. and Koch, K.-F. (2008): *Die Valvetronic: Erfahrungen aus sieben Jahren Großserie und Ausblick in die Zukunft.* MTZ - Motortechnische Zeitschrift, vol. 69(7), pp. 598–605.

[170] v. Basshuysen, R. and Schäfer, F. (eds.) (2012): *Handbuch Verbrennungsmotor. Grundlagen, Komponenten, Systeme, Perspektiven.* 6. Auflage, Vieweg+Teubner Verlag, Wiesbaden, Germany.

[171] Vassallo, A., Cipolla, G., Mallamo, F., Paladini, V., Millo, F. and Mafrici, G (2007): *Transient Correction of Diesel Engine Steady-State Emissions and Fuel Consumption Maps for Vehicle Performance Simulation.* 16th Aachener Kolloquium Fahrzeug und Motorentechnik, Aachen, Germany.

[172] von Grundherr, J., Misch, R. and Wigermo, H. (2009): *Fuel Economy Simulation for the Vehicle Fleet.* ATZ-Automobiltechnische Zeitschrift, vol. 111(00), pp. 2–7.

[173] von Rüden, Kl. (2004): *Beitrag zum Downsizing von Fahrzeug-Ottomotoren.* Ph.D. thesis, Technische Universität Berlin, Berlin, Germany.

[174] Watkins, R. C. (1984): *The Physics of Lubricant Additives.* Phys. Technol., vol. 15, pp. 321–328.

[175] Weinrich, M. W. (2009): *Ein Mittelwertmodell zur Thermomanagementoptimierung von Verbrennungsmotoren*. Expert-Verlag, Renningen, Germany.

[176] Weiss, M., Bonnel, P., Hummel, R., Manfredi, U., Colombo, R., Lanappe, G., Le Lijour, P. and Sculati, M. (2011): *Analyzing on-road emissions of light-duty vehicles with Portable Emission Measurement Systems (PEMS)*. European Commission. JRC scientific and policy reports, (EUR 24697 EN).

[177] Weiss, M., Bonnel, P., Hummel, R. and Steininger, N. (2013): *A complementary emissions test for light-duty vehicles: Assessing the technical feasibility of candidate procedures*. European Commission. JRC scientific and policy reports, (EUR 25572 EN).

[178] Wentworth, J. T. (1974): *Effect of Combustion Chamber Shape and Spark Location on Exhaust Nitric Oxide and Hydrocarbon Emissions*. SAE Technical Paper, (740529).

[179] Wilke, C. R. and Chang, P. (1955): *Correlation of Diffusion Coefficients in Dilute Solutions*. A.I.Ch.E. Journal, vol. 1(2), pp. 264–270.

[180] Woschni, G. (1967): *A Universally Applicable Equation for the Instantaneous Heat Transfer Coefficient in the Internal Combustion Engine*. SAE Technical Paper, (670931).

[181] Wu, K.-C., Hochgreb, S. and Norris, M. G. (1995): *Chemical Kinetic Modeling of Exhaust Hydrocarbon Oxidation*. Combustion and Flame, vol. 100, pp. 193–201.

[182] Yildrim, A. M., Gül, M. Z., Ozatay, E. and Karamangil, I. (2006): *Simulation of Hydrocarbon Emissions from an SI Engine*. SAE World Congress, (2006-01-1196).

[183] Zimmermann, W. and Schmidgall, R. (2007): *Bussysteme in der Fahrzeugtechnik: Protokolle und Standards*. 2. Auflage, Vieweg+Teubner Verlag, Wiesbaden, Germany.

[184] Zschutschke, A., Neumann, J., Linse, D. and Hasse, C. (2016): *A systematic study on the applicability and limits of detailed chemistry based NOx models for simulations of the entire engine operating map of spark-ignition engines*. Applied Thermal Engineering, vol. 98, pp. 910–923.

APPENDIX

A.1 FURTHER RESULTS OF TRANSIENT SIMULATION

A.1.1 *Transient validation of the engine stand-alone model*

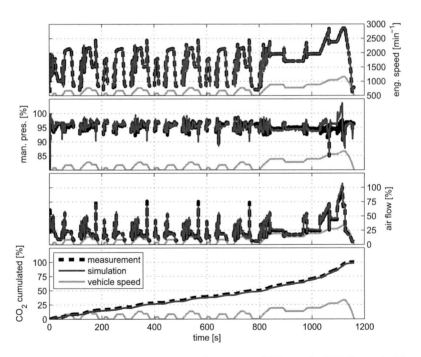

Figure A.1.: Results of several parameters in the engine model during the NEDC started with a cold engine

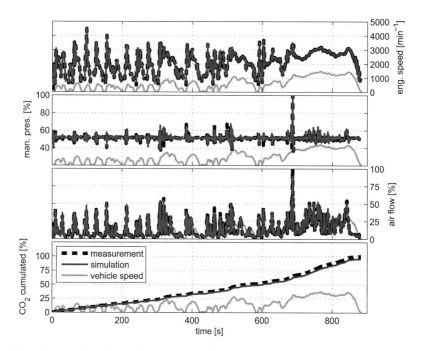

Figure A.2.: Results of several parameters in the engine model during the RCA started with a cold engine

A.1.2 *Transient validation of the driver control model in the coupled simulation*

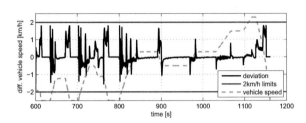

Figure A.3.: Detailed matching of the vehicle speed during the second half of the NEDC started with a cold engine

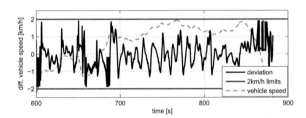

Figure A.4.: Detailed matching of the vehicle speed during the high speed part of the RCA started with a cold engine

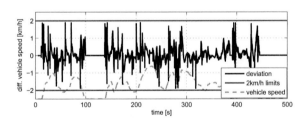

Figure A.5.: Detailed matching of the vehicle speed during the low speed part of the WLTC

A.1.3   *Transient validation of the engine model in the coupled simulation*

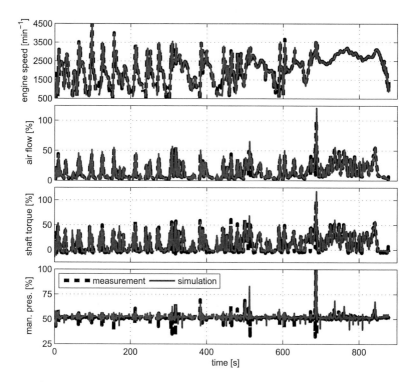

Figure A.6.: Results of the engine model in the coupled simulation during the RCA started with a cold engine

A.2 MODELING APPROACHES FOR THREE-WAY CATALYSTS

source	light-off	oxygen storage	detailed reaction kinetics	pore diffusion	Pt	Pd	Rh	transient	aging
Chatterjee [11]	X		X	X	X		X		
Deutschmann [19]	X		X	X	X		X		
Koop [86]	X		X	X	X			X	
Mladenov [115]	X		X	X	X				X
Koltsakis et al. [85]	X	X			X		X	X	X
Pontikakis et al. [132]		X			X		X	X	
Holder [63]	X	X		X		X	X	X	X
Ramanathan et al. [133]		X				X	X	X	
Tsinoglou et al. [166]		X			X	X	X	X	
Hoebink et al. [62]	X	X	X	X	X		X		
Kwon et al. [90]	X		(X)		X				X
Kang et al. [74]	X		(X)			X			X

Table A.1.: Comparison of several modeling approaches for three-way catalysts arranged to the classification in chapter 3.7.1 [38].

A.3 CHEMICAL REACTION MECHANISM FOR TAILPIPE EMISSIONS

No.	Reaction Equation	ΔH_r	Rate Expression
	oxidation reactions		
1	$CO + 0.5O_2 \rightharpoonup CO_2$	-282.85	$\omega_1 = \frac{k_1 X_{CO} X_{O_2}}{F_1}$
2	$H_2 + 0.5O_2 \rightharpoonup H_2O$	-241.70	$\omega_2 = \frac{k_2 X_{H_2} X_{O_2}}{F_1}$
3	$C_3H_8 + 5O_2 \rightharpoonup 3CO_2 + 4H_2O$	-2042.97	$\omega_3 = \frac{k_3 X_{HC_s} X_{O_2}}{F_1 F_3}$
4	$C_3H_6 + 4.5O_2 \rightharpoonup 3CO_2 + 3H_2O$	-1925.47	$\omega_4 = \frac{k_4 X_{HC_f} X_{O_2}}{F_1}$
5	$CH_4 + 2O_2 \rightharpoonup CO_2 + 2H_2O$	-801.87	$\omega_5 = \frac{k_5 X_{CH_4} X_{O_2}}{F_1}$
	water-gas shift reaction		
6	$CO + H_2O \rightharpoonup CO_2 + H_2$	-41.15	$\omega_6 = \frac{k_6 X_{CO} X_{H_2O}}{F_2}$
	steam reforming reactions		
7	$C_3H_8 + 3H_2O \rightharpoonup 3CO + 7H_2$	497.48	$\omega_7 = \frac{k_7 X_{HC_s} X_{H_2O}}{F_2}$
8	$C_3H_6 + 3H_2O \rightharpoonup 3CO + 6H_2$	373.28	$\omega_8 = \frac{k_8 X_{HC_f} X_{H_2O}}{F_2}$
9	$CH_4 + H_2O \rightharpoonup CO + 3H_2$	206.08	$\omega_9 = \frac{k_9 X_{CH_4} X_{H_2O}}{F_2}$
	reduction of NO		
10	$CO + NO \rightharpoonup CO_2 + 0.5N_2$	-373.25	$\omega_{10} = \frac{k_{10} X_{CO} X_{NO}}{F_2}$
11	$H_2 + NO \rightharpoonup H_2O + 0.5N_2$	-332.10	$\omega_{11} = \frac{k_{11} X_{H_2} X_{NO}}{F_2}$
12	$C_3H_6 + 9NO \rightharpoonup 3H_2O + 3CO_2 + 4.5N_2$	-2739.10	$\omega_{12} = \frac{k_{12} X_{HC_f} X_{NO}}{F_2}$
	dinitrogen monoxide reactions		
13	$H_2 + 2NO \rightharpoonup H_2O + N_2O$	-340.49	$\omega_{13} = \frac{k_{13} X_{H_2} X_{NO}}{F_2}$
14	$N_2O + H_2 \rightharpoonup H_2O + N_2$	-323.71	$\omega_{14} = \frac{k_{14} X_{H_2} X_{N_2O}}{F_2}$
15	$CO + 2NO \rightharpoonup CO_2 + N_2O$	-381.64	$\omega_{15} = \frac{k_{15} X_{CO} X_{NO}}{F_2}$
16	$N_2O + CO \rightharpoonup CO_2 + N_2$	-364.86	$\omega_{16} = \frac{k_{16} X_{CO} X_{N_2O}}{F_2}$
	cerium reactions (heat of adsorption and desorption neglected)		
17	$H_2 + 2CeO_2 \rightharpoonup H_2O + Ce_2O_3$		$\omega_{17} = k_{17} X_{H_2} \theta_{CeO_2}$
18	$H_2O + Ce_2O_3 \rightharpoonup H_2 + 2CeO_2$		$\omega_{18} = k_{18} X_{H_2O} \theta_{Ce_2O_3}$
19	$CO + 2CeO_2 \rightharpoonup CO_2 + Ce_2O_3$		$\omega_{19} = k_{19} X_{CO} \theta_{CeO_2}$
20	$CO_2 + Ce_2O_3 \rightharpoonup CO + 2CeO_2$		$\omega_{20} = k_{20} X_{CO_2} \theta_{Ce_2O_3}$
21	$0.5O_2 + Ce_2O_3 \rightharpoonup 2CeO_2$		$\omega_{21} = k_{21} X_{O_2} \theta_{Ce_2O_3}$

Table A.2.: Chemical reaction mechanism from [63].

The rate expressions are based on total reactor volume and given in $[kmol/m^3 s]$. The reaction enthalpy is defined in $[kJ/mole]$.

The inhibition terms of the reaction rate expressions can be determined as

$$F_1(X, T_s) = T_s \left(1 + K_{a,1} X_{CO}^{0.2} + K_{a,2} X_{HC_f}^{0.7} + K_{a,3} X_{NO}^{0.7}\right)^2 \quad (A.1)$$

$$F_2(X, T_s) = T_s \left(1 + K_{a,4} X_{CO} + K_{a,5} X_{HC_f} + K_{a,6} X_{NO}^{0.7}\right)^2 \quad (A.2)$$

$$F_3(X, T_s) = (1 + K_{a,7} X_{O_2}) \quad (A.3)$$

The kinetic parameter set of the reaction rate coefficients are specified according to

$$k = AT_s^\beta exp\left[-\frac{E}{RT_s}\right], with \tag{A.4}$$

$$A' = \ln A. \tag{A.5}$$

The additional factor for the temperature dependence is neglected ($\beta = 0$) [63].

No.	A' [*]	E [$kJ/mole$]	No.	A' [*]	E [$kJ/mole$]
		reaction	equations		
1	35.60	105.00	12	24.00	80.00
2	36.00	85.00	13	29.80	71.35
3	36.60	112.00	14	31.00	80.00
4	35.30	105.00	15	27.20	80.00
5	28.00	121.00	16	25.80	68.79
6	20.61	67.55	17	7.35	142.30
7	27.70	136.00	18	-6.28	103.09
8	29.90	116.00	19	9.03	142.30
9	26.00	136.00	20	5.38	139.20
10	28.60	80.00	21	-12.02	0.00
11	25.28	71.00			
		inhibition	terms		
1	2.00	-12.75	5	6.80	-3.00
2	6.80	-3.00	6	4.50	-9.90
3	4.50	-9.90	7	11.00	0.00
4	5.77	-12.75			

Table A.3.: Kinetic parameter sets from [63].

A.4 PROCESS SCHEME OF THE OXYGEN STORAGE MODULE

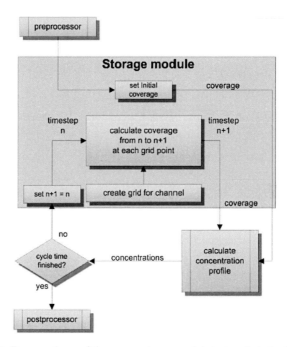

Figure A.7.: Process scheme of the oxygen storage module in transient simulations [38]